Story and Sustainability

M000165796

Story and Sustainability

Planning, Practice, and Possibility for American Cities

edited by Barbara Eckstein and James A. Throgmorton

The MIT Press
Cambridge, Massachusetts
London, England

© 2003 Massachusetts Institute of Technology

All rights reserved. No part of this book may be reproduced in any form by any electronic or mechanical means (including photocopying, recording, or information storage and retrieval) without permission in writing from the publisher.

This book was set in Sabon by Binghamton Valley Composition and was printed on recycled paper and bound in the United States of America.

Library of Congress Cataloging-in-Publication Data

Story and sustainability : planning, practice, and possibility for American cities / edited by Barbara Eckstein and James A. Throgmorton.
    p. cm.
   Includes bibliographical references and index.
   ISBN 0-262-05070-6 (hc. : alk. paper) — ISBN 0-262-55043-1 (pbk. : alk. paper)
    1. City planning—United States. 2. Sociology, Urban—United States. I. Eckstein, Barbara J. II. Throgmorton, James A.
HT165.52 .S76 2003
307.1'216'0973—dc21

                                                       2002043167

10 9 8 7 6 5 4 3 2 1

Contents

Acknowledgments

This book derives in part from a symposium conducted at the University of Iowa in June 2000. Entitled "Planning as Storytelling: Sustaining America's Cities," it was largely funded by a Humanities Symposium Award from the University of Iowa's Obermann Center for Advanced Studies. We thank Jay Semel and his staff at the center, especially Karla Tonella and Lorna Olson, for their able assistance. Additional financial support came from the University of Iowa's College of Liberal Arts and Sciences, the Graduate College, the Office of the Provost, the Department of English, and the Graduate Program in Urban and Regional Planning; and from a grant provided by Humanities Iowa. While money talks, labor makes the wheels go round; special thanks go to Pamela Butler, administrative assistant for the planning program, and to Gayle Sand, administrative assistant for the English department, who together handled all of the symposium's logistic details.

During that event and afterward, we incurred debts to several nonacademic, urban practitioners willing to engage with our project. Despite the press of daily deadlines and the paucity of incentives for contributing to scholarly projects such as this, Joe Barthel, Karin Franklin, Michael Berkshire, and the following people contributed substantial time, energy, and the essential experiential knowledge James Scott calls "metis": Liz Christiansen, administrator of the Iowa Department of Natural Resources' Land Quality and Waste Management Assistance Division; Larry Wilson, campus planner and landscape architect for the University of Iowa; and Jeff Davidson, assistant director of the Iowa City Department of Planning and Community Development. Larry Wilson has been especially gracious, offering insights from his work as a professional planner in Memphis, Tennessee, and Louisville, Kentucky, as well as at the University of Iowa. In addition, we owe a debt to Robert Bullard, whose scholarship has done so much to focus attention on the environmental justice movement. And of course we are grateful to our contributors, who together have brought this volume to life.

In preparing the manuscript for publication, we have learned from the responses of five anonymous readers and our editor at MIT Press, Clay Morgan. This book is better for their advice.

First and last, Barbara Eckstein thanks the fine teachers and aides at Alice's Rainbow Childcare Center. Their inspired work made possible her participation in this and all her scholarly projects.

Contributors

Joe Barthel, Private Investigator, Barthel and Associates Investigations, Oakland, California

Robert A. Beauregard, Milano Graduate School of Management and Urban Policy, New School University

Michael Berkshire, Project Administrator, Chicago Department of Planning and Development

Barbara Eckstein, Department of English, The University of Iowa

Karin Franklin, Director, Department of Planning and Community Development, City of Iowa City

Seymour J. Mandelbaum, City and Regional Planning Program, University of Pennsylvania

Kenneth M. Reardon, Department of City and Regional Planning, Cornell University

Carlo Rotella, English Department, Boston College

Leonie Sandercock, School of Community and Regional Planning, University of British Columbia

Edward W. Soja, Department of Urban Planning, University of California, Los Angeles

James A. Throgmorton, Graduate Program in Urban and Regional Planning, The University of Iowa

Story and Sustainability

Introduction: Blueprint Blues

Barbara Eckstein and James A. Throgmorton

On his first morning in Harlem, Ralph Ellison's invisible man, the central character in his novel of the same name (1952), gazes upon the tall and mysterious skyscrapers far down the island. But he thinks instead about the southern college that he has just left behind. When a man appears beside the curb pushing a cart piled high with rolls of blue paper and singing a blues familiar to the invisible man, he walks along behind, remembering his childhood when he first heard such songs. The man with the cart turns and speaks to the invisible man in jive that uses the rhythms of the city and the folk content of their common southern Negro heritage. Confused yet comforted, the invisible young man follows along "as though they'd walked this way before through other mornings, in other places. . . .

> "What is all that you have there?" I asked, pointing to the rolls of blue paper stacked in the cart.
> "Blueprints, man. Here I got 'bout a hundred pounds of blueprints and I couldn't build nothing!"
> "What are they blueprints for?" I said.
> "Damn if I know—everything. Cities, towns, country clubs. Some just buildings and houses. I got damn near enough to build me a house if I could live in a paper house like they do in Japan. I guess somebody done changed their plans," he added with a laugh. "I asked the man why they getting rid of all this stuff and he said they get in the way so every once in a while they have to throw 'em out to make place for the new plans. Plenty of these ain't never been used, you know."
> "You have quite a lot," I said.
> "Yeah, this ain't all neither. I got a couple loads. There's a day's work right here in this stuff. Folks is always making plans and changing 'em."
> "Yes, that's right," I said, thinking of my letters, "but that's a mistake. You have to stick to the plan."
> He looked at me, suddenly grave. "You kinda young, daddy-o," he said.
> I did not answer. (1952, p. 175)

In his classic postwar novel of urban strife narrated by a once-idealistic, southern Negro migrant, Ellison offers numerous stories that illuminate the challenge of defining, let alone achieving, sustainable American cities. None of those stories more pointedly speaks to the role of urban planners and other professionals in these efforts to imagine the city than this dialogue between the invisible man and the man with the cart full of blueprints that "the man" has discarded to make room for new plans. Thinking of his own very particular plan as a college student, the new migrant insists that "you have to stick to the plan." The man with the cart finds this naïveté sobering.

Ellison does not tell us why the man with the cart retrieves the blueprints or what he does with them; he leaves it to our knowledge of American cities to provide those answers. His man with the cart full of blueprints acts as a metonym for the burden borne by the poorest (chronically, minority) urban residents, who disproportionately have carried the weight of all the social reforms and urban renewal projects meant to manage their lives, and of all the wasteful economies designed to benefit other people elsewhere. While they are the objects of others' very visible and revolving ideas of containment and improvement, those urbanites like the man with the cart have been largely invisible as subjects of their own plans, as builders. Promoters and developers of large urban schemes have thrown away inner-city dwellers and their neighborhoods, making them, literally and disproportionately, the sites of others' trash heaps.

So the content of the blueprints hardly matters to the man with the cart; now discarded, each blueprint is interchangeable with the others. It hardly matters that some were implemented and some were not. What he knows with precision are their collective weight—more than a hundred pounds—and the space they occupy in his cart—two cartloads. Ellison's poor man with the cart and other seasoned observers of American cities will not be surprised that his task is to bear the weight of the proliferating plans and to carry discarded plans out of "the man's" sight.

Nevertheless, the man with the cart is not a sanitation worker assigned a disposal job by contract. Ellison implies that he retrieves the blueprints of his own volition and keeps them somewhere. Perhaps this is Ellison's definition of hope in the inner city: (re)collecting again and again the debris of failed cities, towns, and country clubs, yet imagining that one could truly build something from the wasted material (a paper house in Japan), if only in distant times or places. Singing the blues makes the weight of the discarded plans bearable for the man with the cart in Harlem. However, not even the blues can buoy up his response to the younger man's assumption that a plan is an individually defined goal to which one should adhere in order to progress. For the man with the cart, this reveals a grave misunderstanding of the relationship between plans and America's urban poor, especially its urban poor of color in the twentieth century.

In the course of the novel, Ellison's young student, the invisible man, learns this and other solemn lessons. In fact, faced with many grave problems without good answers, he retreats underground some place in Midtown, where he lives by secretly siphoning off energy from the utility company, Monopolated Light & Power. This clever, transgressive act of gaining energy by fooling the utility company not only keeps his hole warm and light, but also gives significant vitality to the life of the underground invisible man: he is finally his own American "thinker-tinker"—"kin to Ford, Edison, Franklin," as he puts it (1952, p. 7). He is carrying out his own plan. Yet, despite this act of self-determination, his substreet home place remains isolated from that of other thinker-tinkers. And "there is a certain acoustical deadness in his hole" which renders his beloved recording of Louis Armstrong's "Black and Blue" less than it should be (1952, p. 8).

From this hole, the invisible man retrospectively narrates the whole of the novel, the stories of his youth that brought him to the city and sent him underground, where he prepares for some as-yet unrealized future when he can return to street level and act on his own plans. In the beginning of his narrative he states clearly the conditions of his transgressive, underground status and of the hard bargain the circumstances above ground force him to drive: "Irresponsibility is part of my invisibility. . . . Responsibility rests upon recognition, and recognition is a form of agreement" (1952, p. 14). At the end of his narration, he wonders if, even in his continued underground state, he does not have a socially responsible storytelling role to play after all. He asks if "on the lower frequencies, [he does not] speak for you [too]" (1952, p. 581). Only in a more sustainable city, unlike the one he has known, could the invisible man find places in the light of day where he and others could exchange stories on the higher frequencies and enter into trustworthy agreements that form responsible communities.

As editors of this volume, we take up Ellison's implicit call to examine the constitutive role of telling and understanding stories—speaking for oneself and others—in the recognition, agreements, and responsibilities necessary to create and maintain sustainable cities. Specifically, we seek to make space for essays that in varying ways address the interaction of story, sustainability, the (U.S.) city, and democracy.

By assembling this collection of essays, we advance three central arguments concerning that interaction. First, sustainability is both necessary and difficult to pursue. Inherent in the very concept is an endless struggle to achieve an ideal balance of competing legitimate claims for economic growth, environmental health, and social justice. Second, dense urbanized environments raise the human stakes of sustainability. More than 65 percent of people in the United States live in large (a population of more than 500,000) metropolitan areas. Furthermore, the populations of these areas are, on the whole, becoming increasingly diverse racially and ethnically while the economies of these areas are becoming increasingly intertwined with highly competitive, globalized markets. Also, these metropolitan areas use natural resources and produce wastes that move through natural environmental pathways that cross official territorial boundaries. All these facts make the balancing of environmental and public health, social justice, and economic growth through democratic means both more necessary and more difficult to achieve. Third, intensely privatized American cities that emphasize economic growth further heighten the importance and difficulty of pursuing sustainability. This favoring of the private disperses its costs among other people and places, creates enmity, and returns to haunt the beneficiaries of privacy and privilege. In short, story, sustainability, and democracy are mutually constitutive.

The three essays collected in part I—by Barbara Eckstein, James Throgmorton, and Robert Beauregard—advance and refine these basic arguments. Together, they offer a set of ideas about how sustainability, story, and democracy mutually construct one another. By doing so, they also provide themes and tools that readers can use when reading and interpreting the texts that follow.

Distinguishing among authors, stories, and audiences, Eckstein makes three general claims. First, urban planner-authors do not typically reveal how they convert community stories into a single plan or persuasive story, and lacking clarification, readers can rightly respond to such a plan with skepticism. Second, the best stories, most often those that produce a will to change, are those that disrupt habits of thought and defamiliarize conventional expectations. Third, the stories that best serve sustainable cities conscript readers who are willing to engage strangers in dialogue.

In the essay that follows Eckstein's, Throgmorton argues that imagining a sustainable place requires attention to the social and spatial boundaries of *our own* place, and to the importance of unpriced costs, an environmental unconsciousness, tenticular radiations, and remote effects. Progress toward sustainability requires making space for stories that draw attention to a place's ecological footprint, to increasing the efficiency of the value chains that produce that footprint, and to developing a shared sense of moral purpose at a regional scale. Agreeing that a discourse on sustainability includes multiple contending voices, Beauregard argues that storytelling is central to discursive democracy, and discursive democracy, distinct from representative and participatory democracies, is necessary to produce sustainable cities. In a perfect world it would anchor participation and representation and turn attention relentlessly to the need to protect the natural environment, to guide economic growth, and to pursue social justice. However, in an imperfect world such as ours, Beauregard acknowledges, powerful elites mount numerous counterstories that undermine the collective ability to imagine a sustainable city.

Part II of this volume seeks to exemplify, amplify, and modify these key themes and arguments by presenting a deliberate heteroglossia of theoretical and experiential, academic and nonacademic, expository and narrative, familiar and unfamiliar texts. In some cases (Karin Franklin and Michael Berkshire), professional planners narrate tales that come from their practices, tales about the making (and unmaking) of plans. In other cases, community-based practitioners (Kenneth Reardon and Joe Barthel) tell how they assist low-income communities as they resist plans imposed upon them or articulate their own plans "from below." In still other cases, academy-based scholars (Seymour Mandelbaum, Carlo Rotella, Leonie Sandercock, and Edward Soja) offer their interpretations of how storytelling, sustainability, and American cities intertwine. While the diversity of these essays lays out the challenges in the path of sustainability, the seriousness of their subjects also emphasizes the magnitude of the stakes. In other, more academic words, the diversity of these essays calls upon the reader's engagement with the projects of poststructuralism and postmodernism that offer interpretive strategies for negotiating fragmentation, surprise, and uncertainty. At the same time, the seriousness of their subjects urges readers to embrace sustainability as a foundational concept, however protean.

As you consider how the following texts speak *to* you, if not "for you," we encourage you to keep four aspects of story and storytelling in mind. First, storytellers have the power to imagine communities. By constructing relationships between the authors and readers and among readers, they encourage readers to imagine themselves as part of *this* community rather than *that* one, and thereby to recognize the mutuality of responsibility with *these* people rather than *those*. Once imagined in story, such communities then try to defend their claims (for land, for trustworthiness, for material and financial resources, for reductions in risk and uncertainty) against others. Second, story and imagined communities always have a spatial dimension and make a geographic claim. Neither authors nor readers always recognize this spatiality, but it is present nonetheless. Third, storytelling is a constitutive part of democratic practice, and democratic practice is constitutive of sustainability. To serve both sustainability and democracy, stories need to be understood as multiple and listened to with an ear for surprise. And last, the ability to listen with an ear tuned to surprise takes education—formal and informal—in how stories work.

Seeking to develop such an ear, we have organized this book to suggest the repetition, emendation, mutual support, and collective correction necessary to compel the process of sustainable city making. To that end we have juxtaposed the stories and essays in part II so that they speak to one another across disciplinary, professional, spatial, and social boundaries. Short introductions to each piece make explicit the cross-currents among them. Instead of following a structure designed to move the reader from theory to practice or the reverse, we offer an arrangement that moves among story, practice, and theory without favoring any one area.

In the spirit of Ralph Ellison's characters, we are guardedly hopeful. Hopeful that by placing story and storytelling in the foreground, we can make the many social, environmental, and economic consequences of contemporary life in the United States more visible to more kinds of urban users, planners, and scholars. And hopeful that by making these consequences more visible, we can help thinker-tinkers of all types build more sustainable American cities that are responsible to their regions and to the planet they share with others.

I

Storytelling, Sustainability, Democracy, and the American City

Editors' Introduction to Chapter 1

Given the power of traditionally told stories to defend boundaries and galvanize segregated communities, Eckstein argues that planners need analytical tools to reexamine storytelling. To serve the sustainability of cities, storytelling in planning as in art is best understood, she further argues, as an opportunity for a radical turn to disrupt habits and boundaries. An analysis that acknowledges the artifice of stories and thus their potential to disrupt conventional understanding of space and time can flush unexamined habits out in the open.

Social and discursive openness should be understood as both figuratively and literally related to open physical space. Lyn Lofland helps clarify the nuances of "out in the open" in The Public Realm *(1998). Focusing on the connections between social interactions and physical urban design, she emphasizes the distinction not only between private and public space, but also between public and parochial space. While unfamiliar* strangers share public space, *within parochial space familiar strangers see one another time and again engaged in their routines. Parochial space, therefore, may not enable the separation of stories from their function as borderguards. By contrast, what Richard Sennett describes in "The Spaces of Democracy" (1999) as a public realm of public expression active in public space is a place in which unfamiliar strangers can surprise one another with diverse ways of living and narrating those lives.*

Whether the telling and analyzing of stories, particularly in the practice of planning, can, as Eckstein claims, balance the interpretive demands of space and time while taking a disruptive turn is a question that remains open for Edward Soja as he interrogates the possibility in chapter 10 of this collection.

In the public spaces of the city, stories create publics and by creating publics build democracy.

—Robert Beauregard

1

Making Space: Stories in the Practice of Planning

Barbara Eckstein

"Neighborhood is a word that has come to sound like a Valentine," writes Jane Jacobs. "As a sentimental concept, 'neighborhood' is harmful to city planning. . . . Sentimentality plays with sweet intentions in place of good sense" (1961, p. 146). The same could be said of stories and storytelling.[1] Although one all too frequently hears race-baiting, turf-building, boundary-setting stories in buses, bars, coffee shops, and at holiday dinner tables, the belief abides that telling stories, by its very nature, builds tolerant, diverse communities of participant citizens. Even a very clever theorist of story and space like Michel de Certeau does not resist asserting, through metaphor, that "[the story] 'turns' the frontier into a crossing, and the river into a bridge" (1984, p. 128).

Most storytelling—arguably all storytelling—is about setting community boundaries, including some audience members within its territory and excluding others. This boundary setting is inherent in the contract between narrators and audiences that all stories inscribe. Carefully told and carefully heard, stories do have the potential to act as a bridge between engrained habits and new futures, but their ability to act as transformative agents depends upon disciplined scrutiny of their forms and uses. Recently, narrative theorist Peter Brooks worried in the *Chronicle of Higher Education*, that "the very promiscuity of the idea of narrative [in multiple disciplines, in political speeches] may have rendered the concept useless. The proliferation and celebration of the concept of narrative haven't been matched by a concurrent spread of attention to its analysis" (2001, p. B11).

It is attention to this analysis that I mean to offer here. Yet theorists—and with this category of academics I also indict myself—would do well to approach the analysis of storytelling humbled by Seymour Mandelbaum's

reminder that "communities are characteristically indifferent of the proliferation of formal theories. The multiplication of stories, however, puts
them seriously at risk" (1991, p. 210). Just as Jacobs begins her discussion of neighborhoods with a warning about sentimentality precisely
because the safety and success of those urban spaces is so important, I
begin with a warning about sentimentalizing storytelling precisely
because understanding its power is so important to public decision making and urban well-being.

By stories I mean verbal expressions that narrate the unfolding of
events over some passage of time and in some particular location. Stories
use language to frame what has happened to a set of characters in a particular time and place. Although maps, numerical data, computer models, and innumerable other sources of information must be interpreted to
reveal meaning, I do not include those processes of data collection and
interpretation in my definition of story unless that interpretation takes
the form of story as I am defining it. I use this definition first because I
am a literary scholar trained to interpret verbal language in narrative
forms and a humanist attracted to qualitative analysis that acknowledges
the limitations of human knowledge and the uncertainties of life. I also
stipulate this definition in an attempt to distinguish between the modes
of planning theory and practice that rely on quantitative data and computer-generated models to describe structures that set the stage for decision making, and those more marginal modes of planning theory and
practice that attend to continuously unfolding and competing narratives
that create visions of the past, present, and future, and thus impinge on
decision making. This distinction, like most distinctions, blurs under
close examination. However, I think there is often a lot to learn by mind-
ing the gap. So while I have heard scientists tell beautiful, even transformative, stories based on their ability to read numbers or maps, I take to
heart the complaint of Andrew Isserman (1984) repeated in Dowell
Myers and Alicia Kitsuse's omnibus article, "Constructing the Future in
Planning": "planners have adopted quantitative techniques of projection
as if they described the most probable future (truth) and as if that were
desired (ideal)" (2000, p. 224). A focus on storytelling emphasizes the
elusiveness of truth and the complexity of desire. For those who want to
plan for a sustainable urban future, these qualities must be acknowledged and explored.

As a literary scholar who knows that stories are wily and powerful and who cares a great deal about cities and their users, I am compelled to ask, "What can the careful interpretation of stories contribute to the discipline of planning and the future of cities?" This is not a new question for planners.[2] Many have addressed storytelling, writing, and reading as integral elements of planning practice (e.g., Rein and Schon, Schwartz, Forester, Baum, Mandelbaum, Throgmorton, Ferraro, Yanow, Finnegan). In these efforts, some have called upon literary theory, especially narrative theory, as a guide or a bulwark for their interpretive strategies (e.g., Mandelbaum, Ferraro, Yanow, Throgmorton). I would like to revisit this work in my guise as literary scholar and reengage it in a conversation about which questions raised by narrative theory might best serve the needs of theorizing and practicing planners. In the discipline of literature, these questions would be most easily understood if they were sorted into those about authors, those about stories (or texts, if you will), and those about readers or audience. In literature, as in planning, these three categories create and inform one another, often simultaneously. Given this complexity, a solid and familiar tripod erected at the outset may be a useful structure on which to affix those lenses through which we view a necessarily incomplete and unfolding theory of story in planning.

Authors

Although Michel Foucault once declared the author dead—long live the theorist!—publishers continue to display the names and handsome photographs of novelists and poets on their published books. Most readers of literature are confident that they can identify the authors of the books they are reading. The authors of the stories and texts that occupy planners are not so visible. Groups or institutions produce plans, and individual producers often disappear within those groups (Mandelbaum, 1990; Ferraro, 1994). The names of citizens and even city officials whose stories and opinions shaped the plan also do not necessarily appear.

Even if the text at stake is not a plan per se, the question of authorship seems a slippery one. Although diverse and contentious community stories are a given in any place planners are asked to do their work (Mandelbaum, 1990), and planning scholars such as John Forester (1989) and

Howell Baum (1999) offer good advice on how planners—diversely employed—might listen and respond to these stories, the individuals who construct and tell these contending community stories frequently disappear in planning theories about storytelling. Certainly, there are exceptions. Ruth Finnegan's focus on the stories of Milton Keynes (England) residents in *Tales of the City* (1998) is one. However, often such community storytellers are replaced or supplanted by a consideration of planners as storytellers. Perhaps this is because, as Mandelbaum observes (1991), planners convert emotional stories into lists of facts or neutral chronicles in order to contain inflamed passions. Or perhaps it is because planning scholars who care about storytelling are necessarily preoccupied with training the planner as a persuasive storyteller. Whatever the case, it is striking that in Myers and Kitsuse's survey of planning methods that construct the future, storytelling figures prominently and the storytellers are in the end presumed to be planners. "Stories told persuasively can be used to win people over to a planner's way of thinking" (2000, p. 227).

It is certainly true that collaborative authorship can be an ego-reducing endeavor that produces texts that benefit from multiple brains at work. And too, authors need to be good readers and revisers of their own work. It is also true that planners are often hired to act: to collect and process information and offer it to decisionmakers. This said, an amorphous movement between the role of listener-reader and the role of storyteller-author may nonetheless be a hindrance to playing either role especially well. If one listens to others' stories with ears tuned to how their stories will serve one's own storytelling, how they will fit in one's grander narrative, then one risks not hearing them at all. Forester (1989) says as much when he advises planners to listen to people, not simply a line of argument. To this I would add that when planners listen and when they narrate, they too are people who function out of particular social positions and dispositions (Bourdieu, Rpt. 1993, from 1983).

Forester's later ethnographic work (1999) collecting stories of practicing planners seeks to document just such social positions and dispositions (although these words are my imposition from Bourdieu, not his). In Carlo Rotella's recent study of American urban literature, *October Cities* (1998), he asserts that authors are people, a radical assumption in the constructivist world of literary study. This assertion is all the more pertinent in the discipline of planning. It is an assertion enacted in James Throgmorton's study of Chicago's electric future, *Planning as Persuasive Storytelling* (1996) and elaborated in his more recent work (2000).

Identifying the author(s) of a story is the first step in determining who or "what authorizes the author, what creates the authority with which authors authorize" (Bourdieu, Rpt. 1993, from 1980, p. 76). This bit of wordplay from the discipline of cultural study may sound as silly and sinister as Humbert Humbert, the name Vladimir Nabokov gives his pedophilic antihero in the novel *Lolita*, but it states the central question about authors and their moral responsibility to the subject matter of their stories and to their readers. What employers, education, government agencies, degrees, publishers, mentors, universities, gangs, social conventions, and/or means of enforcement legitimate the author, empowering her or him to speak, especially to speak publicly, and be listened to? Mandelbaum describes some of the confusion produced when a reader of a plan cannot identify an author and the associations and institutions that give the author status and legitimacy.

As a citizen-reader I am confused about my own identity when the account of the authorial process doesn't tell me who took initiative, how conflicts were articulated and then resolved, and who if anyone held a veto. At the simplest level, not understanding the author, I can't tell whether the *Plan* is a binding legal prescription or a statement of aspirations. (Mandelbaum, 1990, p. 352)

As a citizen-reader, I too want to be able to identify the authors of the stories planners use and tell so I can assess the bases of their claims to— or, in some cases, presumption of—authority. My goal is not to learn and judge the private details of storytellers' lives, but rather to comprehend their place in systems of power. Only when armed with this information do I have a fair shot at formulating an informed response that stakes a claim for my own authority.

Like Jane Jacobs in her urban neighborhood, I want to establish trust on the basis of shared public concerns. I do not necessarily want to swap divorce stories with the local grocer in my urban neighborhood or the woman next door in my suburban neighborhood. I agree with Jacobs (1961) that too much intimacy places unnecessary demands of private bonds on public trust. In other words, too much intimacy can presume or demand sameness where it does not exist and thus in effect exclude those who are different. We see such a community built on sameness in the 1950s St. Nicholas of Tolentine Chicago neighborhood that Alan Ehrenhalt describes and sometimes celebrates in *The Lost City: The Forgotten Virtues of Community in America* (1995).

It is important to acknowledge the body of scholarship, especially feminist and race-conscious scholarship, that significantly complicates any easy distinction between public and private space or relations. This said, in a democracy a distinction between shareable public concerns and ingroup intimacies that in fact prohibit multigroup public trust is meaningful. I want to do the difficult democratic work of establishing trust based on shared public concerns, not on private individual similarities. I want to know what public associations and convictions inspire and authorize a storyteller.

Pierre Bourdieu complains that in the world of art, the "charismatic ideology" of the individual artist, "the apparent producer" of the poem or painting, actually "suppresses the question of what authorizes the author, what creates the authority with which authors authorize" (1993, p. 76). I am not suggesting that community storytellers or planner storytellers would be better served by replacing anonymity with this sort of charismatic claim to individual genius and public fame. I am instead noting what is at stake when the lines of power that authorize a story or text are erased: trust. Planners and other participants in public discussion who tell persuasive stories to win over listeners without acknowledging who or what authorizes their individual and collective authorship might well be perceived by audiences as marketing agents producing advertisements.

To a humanist like me, public trust is the sacred heart of public policy, so I would like to linger over some examples that can clarify the difficult process of authorizing authors. I am intrigued by practicing planner William Lamont's (1990) impatient response to Mandelbaum's "Reading Plans" (1990). In it, he disregards Mandelbaum's focus on citizens' reading of a final plan and their confusion about authorship. Instead he returns to the writing of a plan and notes, in reply to Mandelbaum's criticisms, "We [planners for a central Denver plan] shared responsibility for content with 100 citizens (many of whom were urban development professionals). We also relied on citizens to challenge the ideas in the plan for their defensibility, reality, and clarity. Once the plan has been adopted the important thing is to carry out the actions and to monitor the results.

The report itself is history" (Lamont, 1990, p. 358). While this process gives consulted citizens an enormous responsibility, it asks them to carry out that responsibility in response to a draft plan that then leaves their hands, takes final form through someone else's effort and decision making, and reaches the general public as "history," that is, it seems, no longer amenable to reading, interpretation, and revision by citizens. And yet actions are to ensue from this final form.

Designated citizens are asked, in the plan-writing process, to challenge the reality of the plan with their own stories of reality, but they know that those stories as told by their authors will not appear and indeed that all of their challenges will be managed by a group of planner-authors who will decide on the public presentation of reality. In the chasm between consultation and history, the agency of those consulted citizens disappears; their stories leave their control; and the citizenry at large have no agency at all. Yet the planning authors (or Lamont speaking for them) can announce that the citizen representatives have authorized the plan.

My own experience as an early consultant for my university's strategic plan has taught me how debilitating a role this is. I felt the heavy weight of the responsibility Lamont describes. I was called upon to disarm faculty skepticism and generate faculty enthusiasm; to represent other faculty, students, and staff; and to imagine the best future for the institution. Still, I was keen to participate. But I soon learned that every word I uttered or wrote would be transformed by a committee of the president and vice presidents into a document containing the data they believed would be most attractive to the Board of Regents, the state legislators, major donors, major granting agencies, and perhaps other audiences of which I was unaware.

Ironically, my considerable faith in the president and vice presidents as ethical individuals with whom I had worked successfully evaporated in this process, which defaced my contributions while it used them to underwrite the authority of the document. If university administrators had simply collected information from their data gatherers and then written the political document they believed they needed, the authorship and the authorization of authorship of that document would have been clear. As it was, the exploitation of citizen participants only produced cynicism. Mind you, I would have preferred a participatory process that actually granted power to those consulted, but this would have been an utterly different exercise from the one in which I was asked to engage.

Given that the present conventions of plan writing rarely call for identification and authorization of authors, and that much of the storytelling in which planners engage becomes unmoored from its authorship, it is reasonable that Mandelbaum and Throgmorton turn to the concept of ideal author to describe the authorship of planning texts (Mandelbaum, 1990; Throgmorton, 1996). Ideal authors are created by the texts themselves; they are the authoritative, confident voices and dispassionate visionaries that speak out of the text, untroubled by careers or other vicissitudes of life. And they create ideal readers who meet them on their own terms (Mandelbaum, 1990).

Although inevitably authors do imagine readers when they write and readers imagine authors when they read, this process is, I would argue, best kept tied to geohistorical realities. I address this question in my discussion of readers, but voice here my concern that the concept of ideal author not supplant the interrogation of real authors and the powers behind them. Mandelbaum (1990) explains that through indirect sources such as newspapers he fills in the information about authors and characters missing from a plan. Throgmorton (1996, 2000) more thoroughly describes all relevant characters-agents in his work. This seems to me a more important interpretive strategy than the construction of an ideal author.

Whether planners or urban users, storytellers have intentions that they mean to promote through whatever powers rise to support their narratives. When public decisions are afoot, every storyteller is narrating to control the actions of others. Stories that can be acted on are necessarily the ones valued in planning (Rein and Schon, 1977). However, the link between the intentions of authors and the actions of audiences is hard to forge and harder to measure. William Baer (1997) does offer evaluation criteria for a good plan that attends to the stages of the planning process. He also addresses the postmodernist concern with the multiple voices in a community by recommending that "the plan should be in a form suitable for debate" (1997, p. 341; relying on Kent, 1964). A platform for debate is a substantially more hospitable form than "history," door closed. Still, the whole network of stories received and told in response to a certain planning issue, moment, and territory is bound to be in excess of any author's intention to contain it. In the world of literary interpretation, the New Critics of the mid-twentieth century threw up their political hands and declared that all discussion of intention was a fallacy. "The poem is not the critic's own and not the author's (it is detached from the author at birth and goes about the world beyond his power to intend about it or control it)" (Wimsatt and Beardsley, 1954, p. 5). They are equally skeptical about the ability to measure a text's effect on its audience and refer to this critical endeavor as the affective fallacy. While their warnings have merit, even in planning, their conclusions are extreme.

What one can say with certainty, I think, is that examining authors' intentions without also investigating their effects on audiences is of no political use. In literary circles, scholars still argue about whether a very intentional novel such as Mike Gold's communist-influenced, Popular Front book, *Jews without Money* (1935), actually reaches more readers and more classes of readers and provokes them to act as Gold intended than does a convoluted form such as Henry Roth's *Call It Sleep* (Rpt. 1964, from 1934), a Joycean-influenced novel of modern urban dangers and delights that is also, like Gold's novel, about early twentieth-century New York City. Gold's style could hardly be more direct or his intention more obvious; Roth's could hardly be more indirect or more challenging to unveil. Which has been more read and more provocative to urban users? And to which urban users? Luckily for planners, the destination of plans and the other stories planners hear and tell is less diffuse than the destination of published novels. A circumscribed geography and time frame may make the execution and reception of intention more measurable, as Baer suggests. However, I still read such public intentions with a skepticism learned from the story of Mike Gold.

How then are we to understand the role of the author? The Inuits say that the storyteller is the one who makes space for the story to be heard. As I see it, this definition of the traditional storyteller cum author may be especially useful for planners, for it assumes that the stories storytellers tell are not their own. It is their knowledge of traditional stories and local conventions; it is their skill as narrators, as "hosts," for stories they hear and retell; it is their demeanor, their voice, their ordering, their shaping, their ability—literally—to create an amiable narrative and physical space, that allow their telling, retelling, and thus transformation of the community's stories to be heard.

In such a space in the present, the aggregate and arrangement of past stories can tell the community's future. In this way, stories can be the bridge de Certeau describes. In this sense of making space for others' stories to be heard, the planner is a modern embodiment of the storyteller in traditional societies. Seen this way, planners' work might then be judged by how successfully they create a three-dimensional or textual space amenable to multiple stories, how well the arrangement of that space produces provocative interaction among the stories, and thus how well and how broadly the stories are heard.

However much this present space of storytelling is squeezed by stories of the past and plans for the future, it should not be permitted to disappear between them, as Mandelbaum (1984) suggests that it does. It is in the present that votes are cast, decisions made. The influential metafiction writer Jorge Luis Borges is right to show us, in his well-known story, "The Garden of Forking Paths" (1964), that decision making is not often a clearly forking path, that present decisions cannot altogether determine the future or relinquish the past. That political present, that fork in the path, is nonetheless crucial to keep in view. It is the storyteller's job to make space for that present and the presence of multiple stories in it.

Stories

Just what constitutes a story in the practice of planning is not an easy question to answer. Plans are texts and contain narratives, but from my outsider's perspective, I take it that they are not necessarily the goal of planning. Even when they are produced, their forms vary so much according to the dictates of government regulations and employers that it is difficult to place them in any one genre. If the planner—the postmodernist planner—attends to the multiple voices of the community, one cannot expect a plan to be other than contingent fragments (Beauregard, 1991). In the judgment of Giovanni Ferraro, "planning texts tend to be ephemeral and evanescent: one could say that they hardly meet the condition of being a *chose du text*, independent of their authors" (1994, p. 210). Yet he goes on to write of them as though they are possible subjects of textual analysis. The complications of plans notwithstanding, planners tell and hear a great many stories in the course of their work, so the concept of story is a salient subject of planning literature.

Planning scholars who draw on stories for their theories reasonably assume that a central value of stories is their ability to bring order to the chaos of events (Rein and Schon, 1977; Mandelbaum, 1990; Yanow, 1995; Throgmorton, 1996; Schwartz, 1996; Myers and Kitsuse, 2000). For Peter Schwartz, "stories have a psychological impact that graphs and equations lack. Stories are about meaning; they help explain why things could happen in a certain way. They give order and meaning to events— a crucial aspect of understanding the future possibilities" (1996, p. 38; quoted in Myers and Kitsuse 2000, p. 227). Addressing problem setting in particular, Myers and Kitsuse assert that "storytelling illuminates the whole of a problem by forcing problem setters to identify the key actors and the chain of events that lead to the circumstance perceived as problematic" (2000, p. 229). They further assert that the validity of stories is measurable and offer five criteria from Rein and Schon to test this validity: stories should be consistent, testable by empirical means, lead to a moral position, be capable of being acted on, and be beautiful (Rein and Schon, 1977, p. 5; quoted in Myers and Kitsuse, 2000, p. 229).

Some planning scholars who write about stories call upon narrative theory to ground their arguments (Mandelbaum, Yanow, Throgmorton, Ferraro), and most of those arguments are for the order stories bring to life events. Ferraro is an exception in that he relies on Roland Barthes's *Le Plaisir du Text* (1973) to suggest that planning texts be more radical by being more playful. Attractive as this line of reasoning is to a literary scholar, I doubt that citizens contemplating a freeway through their favorite park are much in the mood for play. Dvora Yanow's analysis of "Built Space as Story" (1995) is a more typical use of narrative theory. She cites Tzvetan Todorov, who declares that "events don't tell themselves" (1981, p. 39), a statement that served E. O'Connor in "Telling Decisions" as well (1996). She also quotes Hayden White: narrative is a "form of human comprehension that is productive of meaning by its imposition of a certain formal coherence on a virtual chaos of events" (1981, p. 251).

Noting that White says "a certain formal coherence" and "a virtual chaos of events," I would like to suggest that this presumed paradigm of stories bringing order to a chaos of events can also be reversed. Stories can quite usefully disrupt the habits of thought and action that control everyday life. Although many traditional (turf-bound) stories do maintain order in a given community (or nation, even) and an over-arching narrative offered by a persuasive voice can articulate a linear cause and effect that produces a representation of order in contentious circumstances, many practices of everyday life are in themselves already habitual. These habits offer a powerful semblance of control, containment, reason, and safety that is often necessary to live day to day but which inhibits innovative change. We walk against a traffic light at our peril. We sleep on the "wrong" side of the bed at risk of a restless night. We fear and ignore homeless people because we cannot fit them into the routines of our kitchens and offices. The healthy middle and upper classes can better afford order than those who are unwell and/or poor, to be sure, but even those without dependable bodies or stable homes assert whatever order, follow whatever habits, they can muster. Life is lived as habit and succumbs to chaos only when there is no other choice.

All stories do select certain events to tell and others to ignore, but this "certain formal coherence" might not necessarily bring narrated linear order to lived chaos. It could instead take the habits of everyday life and renarrate them in an unfamiliar and thus usefully disruptive way.

By modern standards, the best art does just this. It defamiliarizes the everyday, encouraging its audience to rethink their humanity and their place in society. It seems to me this is what Howell Baum is moving toward in "Forgetting to Plan" (1999). Unless a neighborhood's habitual, traditional stories are acknowledged and transplanted, plans for the future of the present, diverse neighborhood cannot look different from the remembered, homogeneous past.

"Stories diversify; rumors totalize," writes de Certeau (1984, p. 107). Not all stories diversify by any means, but stories do have this potential, and to the extent this potential is exploited, de Certeau's distinction is useful. The will to change has to come from an ability—a planner's ability, an urban user's ability—to imagine one's self in a different skin, a different story, and a different place and then desire this new self and place that one sees. It has to come from a storyteller's ability to make a narrative and physical space in which to juxtapose multiple, traditional stories so that they enrich, renarrate, and transform that space rather than compete for ultimate control of a single, linear, temporal history of an impermeably bounded geopolitical place.

How can planners detect the defamiliarizing elements in a story they hear or construct a story that defamiliarizes? The simple answer must be mindful practice; the full answer is in what that practice reveals. Donald Schon's *Reflective Practitioner* (1984) describes a mindful practice in which one learns by doing; John Forester's *Deliberative Practitioner* (1999) adds to this prescription the need to reflect in dialogue with others. My understanding of defamiliarizing stories is amenable to these well-wrought arguments for a reflective and dialogic practice, but I mean to prod the deliberative practitioner to be deliberate as well.

To planners, the word "deliberate" may connote modernist top-down planning and desires to stream-line and simplify communal complexity for the sake of comprehensive plans and immediate action. So I hasten to stipulate that I am suggesting that planners be deliberately armed with explicit knowledge about how stories work as narrative forms. With this knowledge they can recognize and enhance defamiliarizing possibilities in stories whose forms are improvisations of traditional, habitual patterns of life and narration.

Stories manipulate time, voice, and space to produce forms that are intended to be compelling. And they can be most compelling, I argue, if they prepare us for and produce ruptures in our quotidian lives. The structuralist Gerard Gennette (1971) has devoted his career to mapping the manipulations of time in literary fiction, paying particular attention to the differences between the duration of events and the duration of the narration of those events, and to the frequency of repetition of events and that frequency in stories about them. In the planning profession, where practitioners are employed to measure the circumstances of the past in the present in order to plan for the future, it might be especially useful to attend to manipulations of duration and frequency.

Examples employing duration and frequency, both in and out of literature, will, I hope, clarify their meaning. In Ann Petry's *The Street* (1946), a novel of 1940s Harlem, a third of the novel's pages narrate the events of one night when a young, single mother leaves her small apartment and her 8-year-old son to have a beer in a corner bar. Far fewer pages are devoted to the years of her marriage, her domestic employment in Connecticut, which precipitated the end of her marriage, and her subsequent residence in Harlem. The speed with which Petry tells that familiar story is balanced against the slow pace at which a single evening moves past. Through the course of that evening the reader is made to dwell at length in the dilemma of a poor, single mother and leave the easy chair of social theorizing about poor, single mothers, and especially single, inner-city, African-American mothers. The speed with which tragedy ensues in the pages depicting the days after this fateful evening confirms the reader's suspicion that the conundrum of that one night is the crux of the story we are given to understand. Attention to duration in storytelling allows one to hear what matters most to storytellers and perhaps even to learn why it matters.

Skill with duration in storytelling affords one the opportunity to rivet attention on those events and occasions that best serve one's intention. To be sure, one may manipulate duration, however unwittingly, to hide what one chooses not to reveal. Formal knowledge of storytelling cannot guarantee moral storytelling. However, interpreters informed about the ways stories work can collectively push storytelling toward moral practices and ethical ends.

Like duration, frequency of repetition always matters in interpreting life and story. Whether a complaint, an image, or a musical interval, frequency of repetition produces patterns of significance. A literary example that may resonate with planners is one in which the same events are narrated by different voices. One famous novel using this technique is William Faulkner's *The Sound and the Fury* (Rpt. 1956, from 1929). Faulkner not only repeats in different voices the events marking the demise of a plantation family, he arranges those voices from the least comprehensible to the most comprehensible. The novel begins with Faulkner's representation of the voice of a severely mentally retarded adult. He risks alienating readers in order to disorient them sufficiently so that when, finally, they find a comprehensible safe haven with a narration controlled by the central consciousness of the family's black servant, they fully appreciate the singular sanity of that character.

This modernist experimentation may seem substantially beyond what the traffic will bear in planning circles. However, if my own political, social, and professional experience is any measure of how often this technique is used outside of literary texts, then Faulkner is not alone in this method of instruction through defamiliarization and repetition. I think here of my forays into the collection of oral histories and of those who wanted to judge whether I would risk a humbling disorientation in order to learn what they had to teach me. Once, for example, when interviewing a Lakota elder in Chicago about her movement to and within the city, I never got an answer to my initial question about using a tape recorder—a simple, straightforward question, I had thought. Seemingly deaf to my repeated question, my hostess offered me store-bought cookies and an endless unfolding of stories, all of them only tangentially related to what I thought was our agreed-upon topic. Slowly I realized I was witnessing a provocative analysis that never left the mode of narration or the manner of cordial indirection. Only after we had left the Indian Center and were parting on a noisy Northside street corner did she tell me a story directly relevant to my questions. Whether the disorientation comes in the form of evasion; rapid-fire slang; dense, professional jargon; or direct insults, I have learned that the repetition of the story at stake—often in multiple voices, often beginning with the least comprehensible—teaches me to hear it. First, however, I have to relinquish habits familiar to me.

Having learned the pain and value of disorientation through reading stories and attempts at collecting oral histories, I am grateful to find my conclusions elegantly corroborated by the oral historian Alessandro Portelli (1991). He writes of an inter/view (an invitation to storytelling) as a mutual sighting of two subjects and notes the particular challenges of such mutuality in encounters between members of the professional elite and members of "oppressed or marginal social groups." "Not only the observed, but also the observer is diminished and alienated when social conditions make equality impossible" (1991, p. 32). Portelli's countryman and fellow scholar-activist, Ernesto de Martino, explains,

Reopening a dialogue between two human worlds which long ago ceased to speak to each other is a difficult enterprise, and it causes many burning humiliations. It humiliates me to be forced to treat people my own age, citizens of my own country [also city], as objects of scientific research, almost of experimentation. It humiliates me when they take me—as they have—for a revenue agent or for a show business entrepreneur. (1953, translated in Portelli 1991, pp. 31–32)

Portelli concludes that the interview cannot create equality where it does not exist, even as the very process demands such equality. To address this conundrum of necessary, disorienting humiliation on all sides, he recommends making the interview an experiment in equality by dealing with power openly. For scholars and other academically trained professionals this means, I think, removing the armor of knowledge-based competence designed to deflect any public humiliations and developing an ear for the rhythms of duration and frequency in others' stories.

In the narrative theory of Gennette's generation, questions of space recede behind these questions of time. However, in the ensuing decades, space has appeared as an equal partner in the interpretation of events, whether historical or fictional (Bakhtin, 1981; de Certeau, 1984; Soja, 1989). Both the spaces made for storytelling and the spaces stories make figure in the production and apprehension of meaning. I discussed the first as a key element in the definition of storyteller and will return to this issue of spaces made for storytelling in my treatment of audience. I focus here on the spaces that stories make.

In the territory of narrative theory, at the crossroads of time and space, sits Mikhail Bakhtin's idea of the chronotope, literally time-space. First developed by Bakhtin in the 1930s from a term used in biology, the idea became widely available to English-reading audiences in 1981. Assuming the "intrinsic connectedness of temporal and spatial relationships" (1981, p. 84), Bakhtin identifies motifs or images—chronotopes—that represent such interconnectedness and thus constitute the form of various genres of narrative through western history: the road of Greek adventure tales, the castle of the gothic novel, the salons of nineteenth-century French naturalism.

Contemporary scholars, among them those intrigued by the virtual reality of the Internet and the "space" of hypertext, have found rich provocation in Bakhtin's theory of the chronotope. It is no less provocative for urban scholars and practitioners who must, with the acuity of Balzac, " 'see' time in space" (Bakhtin, 1981, p. 247). The public meeting room, the landfill, the council chambers, the neighborhood, the yard—these and many other "chronotopes [are] formally constitutive categories" of stories that planners hear and tell (Bakhtin, 1981, p. 84).

The arguments of geographer Edward Soja (1989) encourage close attention to geographic scale as another important factor in the production of meaning. Stories use different geographic scales and their interpretation depends upon careful reading of those scales. For example, while chemists employed by corporations or the U.S. Environmental Protection Agency can tell a story on the scale of the molecule or on the scale of parts per million of contaminants in an entire river system, parents, such as those in Woburn, Massachusetts, tell a story on the scale of the human body; specifically, the bodies of their leukemia-afflicted children who drank the water from the river. A doctor who can draw a circle around a geographic region on a map and see a cancer cluster is reading on yet another scale (Harr, 1996).

In addition to scale, spatial perspective matters. Don't look down on the city from the skyscrapers, say Lewis Mumford, Jane Jacobs, and de Certeau; interpret from the sidewalk. The same space will be a different place. An important corollary to familiar assertions about the importance of spatial perspective is bioregionalists' attention to the issue of remoteness. Ecofeminist Val Plumwood's (1998) explication of remoteness provides an especially rich understanding. She challenges earlier bioregionalists' argument that small-scale autarchic communities are the ones best designed to recognize and counter adverse ecological relationships. She asserts instead that small-scale self-sufficiency is "neither necessary nor sufficient to guarantee that other important forms of remoteness are avoided" (1998, p. 567). The other forms of remoteness she identifies are consequential remoteness (where the consequences fall systematically on some other person or group, leaving the originator unaffected), communicative and epistemic remoteness (where there is poor or blocked communication with those affected, which weakens knowledge and motivation about ecological relationships), and temporal remoteness (from the effect of decisions on the future).[3]

Plumwood concludes that reduction of remoteness needs to be investigated as a political and not just a spatial organizing principle. "The whole thing is totally wrong," remarks LeAlan Jones, teenager and neighbor of the Ida B. Wells housing development. "The University of Chicago is walking distance from the Ida B. Wells. It has some of the greatest Nobel Prize winners of this last century, nuclear physicists that almost cracked the atom—but kids around here don't even know it's there. Some kids have never even been downtown" (Jones and Newman, 1997, p. 153).

Stories that defamiliarize can compel audiences to shift their usual interpretive scale or spatial perspective. This is not to say that audiences, stakeholders, easily give up the scale of their own backyards when a possible landfill is close at hand. In the course of practice, planners surely hear and rehear many of the same stories and observe the predictable responses of their audiences, which seem as programmed as the identification and desire produced by grocery store romance novels. The not-in-my-backyard (NIMBY) story in response to locally unwanted land use (LULU) is so familiar that in defense of their own sanity, planners (that is to say, all those employed to negotiate the LULU on behalf of the public at large) have reduced the tale to a Swiftian vocabulary of humor and frustration. But of course the very repetition of the story speaks to its enormous importance and the stakes in breaking the habits of all tellers of and listeners to this tale. Planners—indeed any participants in deliberations who are inured to this story—stand a chance of usefully disrupting this narrative if they listen deliberately to the form in which it is told, pursuing the opportune moment and the material for a defamiliarizing turn.

In collecting stories about the 1949 death of Italian laborer Luigi Trastulli, Portelli (1991) learns not only how often subsequent generations of laborers in Trastulli's home region repeat the story, but also how frequently the stories contain factual errors, most notably that the shooting by police took place in 1953, not 1949.[4] Rather than seek factual truth in people's memories and then dismiss their stories when that memory strays from documented fact, Portelli looks specifically for emotional truth in the oral narratives, truth most pointedly revealed in the production and repetition of errors. This is to say that hope for sustainability, whether it means preservation or change, may reside in a planner's ability to distinguish story truth from data truth and to recognize, interpret, and defamiliarize the use of duration, frequency of repetition, voice, chronotope, scale, spatial perspective, and remoteness in the stories they hear and tell. De Certeau defines narration as "having a content, but also belonging to the art of making a *coup* (hit): it is a detour by way of a past ('the other day,' 'in olden days') or by way of a quotation (a 'saying,' a proverb) made in order to take advantage of an occasion and to modify an equilibrium by taking it by surprise" (1984, p. 79). I would challenge planners—all of us who care about urban futures—to make space for NIMBY stories, attending to their detours, quotations, and errors, so that we learn the art of fruitful surprise.

Readers and Audience

Although identifying one's audience before the story is made and told may be the best hope of creating a story that succeeds in linking one's intentions to desired effects, such identification is not always possible. (Outside of writing workshops, it is never possible in literary production. In planning, the odds are a bit better.) Mandelbaum (1990) and Throgmorton (1996) both suggest that the best way around this dilemma is to recognize that the rhetoric of a text inevitably creates an ideal author who in turn produces an ideal reader, that citizen who will act as the author-planner would have him or her act.

To retain the fact of inscribed authors and readers but avoid the complications and remoteness of ideals, the inscription of the audience might instead be thought of in the terms provided by narrative theorist Garrett Stewart. In his study of nineteenth-century British fiction (1996), he defines what he calls the conscripted reader. For Stewart the word "conscripted" combines the fact of readers' roles being inscribed in a story with the idea that a text drafts readers, however voluntarily, to play particular roles and embrace particular beliefs and values. "Implicated by apostrophe or by proxy, by address or by dramatized scenes of reading, you are deliberately drafted by the text, written *with*. In the closed circuit of conscripted response, your input is a predigested function of the text's output—digested in advance by rhetorical mention or by narrative episode" (1996, p. 8). Stories of planners and urban users may not address their audiences with the explicit civility of the Victorian novel's Dear Reader, but such stories do attempt to enlist their audiences to feel and think and act as the story proposes.

A direct address to and conscription of readers that is not so far from that in Victorian novels is evident in Andres Duany, Elizabeth Plater-Zyberk, and Jeff Speck's *Suburban Nation* (2000). Its introduction begins,

You're stuck in traffic again.

As you creep along a highway that was widened just three years ago, you pass that awful billboard: COMING SOON: NEW HOMES! . . .

Those one hundred acres, where you hiked and sledded as a child, are now zoned for single-family housing. . . .

It is not just sentimental attachment to an old sledding hill that has you upset. It is the expectation, based upon decades of experience, that what will be built here you will detest. . . .

You are against growth, because you believe that it will make your life worse. And you are correct in that belief, because, for the past fifty years, we Americans have been building a national landscape that is largely devoid of places worth caring about. (2000, pp. ix–x)

Such conscription is not the end of the reception process. Actual readers—what I will call geohistorical readers—negotiate with the conscription in accordance with their interpretive communities (groups determined by cultural and professional training or practice, e.g., English professors, electricians, hog producers, inside traders), the formative experiences of their individual lives situated in a particular time and place, and their dispositions. Those forces shaping the actual reader may in fact be so far outside of the conscription proposed by the story that the reading is a blatantly resistant one. Resistant gay interpretations of Westerns, for example, have substantially changed the way that many viewers perceive that genre, which was assumed for years to be a bastion of heterosexist masculinity.

I have a resistant reading of the introduction to *Suburban Nation*. As soon as I read, "You're stuck in traffic again," Duany et al. have lost me as a conscriptee because I have lived my life in strict avoidance of auto traffic jams. I will walk a mile—two—to work or wait with equanimity for a bus, train, or plane, but I will not take a job, home, or vacation that puts me in an auto traffic jam. My stubborn resistance to this first conscripting sentence has everything to do with my subsequent skepticism about the whole vision of the American dream held by Duany et al. Their bridge—or freeway overpass—between "you" and "we" is not one I am on. Even though I agree that "we Americans have been building a national landscape that is largely devoid of places worth caring about," this is the case because "you" qua "we" are willing to participate daily in freeway sprawl. Of the many Americans who *are* caught daily in traffic jams, some will enlist in this story as they are conscripted to do. Out of their interpretive communities, social positions, and individual dispositions, they will tell stories that elaborate and corroborate the story of hiking and sledding told for them in the introduction to *Suburban Nation*. Some will literally invest in the vision of *Suburban Nation*. Meanwhile, resistant readers of *Suburban Nation* together reform and renew its vision as queer interpreters of Westerns have revitalized that foundational American drama.

The title of *Our America* describes a conceivable space shared by "we" who "live in a second America where the laws of the land don't apply and the laws of the street do" and "you" who "must learn our America as we must learn your America, so that maybe, someday, we can become one" (Jones and Newman, 1997, p. 7). These words of LeAlan Jones sit on the opening page of the book above the lyrics of a gospel hymn that begins, "Do you remember when you walked among men?/ Well, Jesus, you know as you're looking below that it's worse now than then." The conscripted "you" who does not know and needs to learn about the lives of two African-American men growing up in the ghetto is juxtaposed to the direct address of "you," Jesus, who also needs to be educated about the deteriorated condition of human existence. The voice of the hymn trusts that Jesus, once informed, can "show me the way. One day at a time." This powerful juxtaposition of the two "yous" implicitly challenges the reader to live up to the example of Jesus. While the reader cannot be in a position to show the way, she or he is conscripted as educable, as one who can use the means available to act on what is learned in these pages. The book ends with Jones's wishful transformation of "I" and "you" into an inclusive, active "we."

I know you don't want to hear about the pain and suffering that goes on in "that" part of the city. . . . But little do you know that "that" part of the city is your part of the city too. This is our neighborhood, this is our city, and this is our America. . . . We've got to make a change. . . . Not me by myself. Not you by yourself. I'm talking about all of us as one, living together in our America. (Jones and Newman, 1997, p. 200)

As an actual reader, I am moved by a teenager's trust that the book's conscription offers me and humbled by the depth of the problems facing us in our neighborhood of our city in our America (in our world). Yet I wonder how "we" on the freeway of a suburban nation will get this message, which is not posted on a billboard along the highway.[5]

The use of the second-person "you" or even first-person plural "we" is not the only way to define the reader and his or her task. In *Planning as Persuasive Storytelling*, Throgmorton (1996) conscripts the reader by presenting himself as an "I" situated in many different and specific times and spaces, an eye who sees (and hears) for the reader. It is, to a literary eye, the strategy of Walt Whitman, who used his own capacious presence and voice to "contain multitudes," to make space for all classes and breeds of Americans and therefore for a participatory democracy on an inclusive model. Whitman's is a voice that sets out to heal the wounds of the American Civil War. In Throgmorton's meticulous attention to the many people and places, maps and charts, advertisements and news reports, arguments and humiliations that made up the struggles in late 1980s over the electric rate hikes and nuclear power plants in the greater Chicago area, his senses, like Whitman's, stand in for the reader's. This nonlinear strategy is hard to capture in short excerpts, but two will suggest the method.

The CEOC's [Chicago Electric Options Campaign] building sits near a busy intersection. It is a one-story yellow stucco building with newer windows. Cars and buses roar by, spewing exhaust and kicking up dust. . . . I walk down to the basement, where the office of the CEOC's one staff member is located. It all looks very temporary, but with ample room for a large number of people to gather. A place for organizing. (Throgmorton, 1996, p. 173)

. . .

I walk down a dreary hallway, pass by the press room, and come to the alderman's committee office. . . . The phone rings frequently. . . . People drop by constantly: Betty to talk about zoning for Children's Hospital, former campaign workers to chat, a friend from the press office to use the computer, John Hooker (Com Ed's lobbyist) to talk about a transmission line, Maureen Dolan from the CEOC to talk about the franchise and an energy conservation hearing, a fellow alderman who refers to CEOC people as kooks. (Throgmorton, 1996, p. 193)

The reader is conscripted to trust the acuity of these observations and draw political and moral conclusions accordingly.

This defamiliarizing narrative strategy (which is particularly unfamiliar in the discipline of planning) is persuasive storytelling for this particular geohistorical reader because it provides me a credible and comprehensible wealth of detail about Chicago's electric dilemma: both conventional information from the esoteric world of utility planning and unconventional information about the settings and characters that staged the controversy. My public trust is enlisted by the "I," which has researched, witnessed, watched, and crafted an important and complex set of utility decisions that I would otherwise be forced to ignore because of their remoteness from bodies of knowledge and modes of expression that I can interpret. Throgmorton conscripts a reader capable of understanding an arcane industry and sharing in a convoluted political process that shaped the future in ways that affected all urban users in the Chicago area and that touches all utility users in America.

Throgmorton also conscripts a reader whose public trust of the "I" will be enhanced by the candid revelation of personal details of his life. In his later work, Throgmorton (2000) presents an even more self-revealing, personal "I." This candor enlists a reader who distrusts judgments that derive from remote expertise and who prefers a knowledgeable narrator and planner walking amidst all stakeholders and agents in the drama. While as an actual reader I sign on for this definition of dispersed agency and expertise, I resist the text's suggestion that I corroborate the "I"'s particular inscribed emotions in response to life's exigencies. This is more togetherness than I need in order to trust that this speaking "I" has a keen eye for virtually all the significant details of the public issue at stake.

I think again of Jane Jacobs (1961) arguing that suburbs fail to be lively, safe, and diverse neighborhoods because they demand too much togetherness on private ground as a result of providing no public spaces in which diverse people can meet amiably as strangers or acquaintances. People are a lot more selective about who they invite into their kitchens than in who they shop next to at the drugstore or sit next to in a bar or agree with about a school board candidate. Throgmorton's method provides ample public space in which diverse readers can meet. Making space in trains, on streets, in mayors' offices, court rooms, council chambers, and in his own texts for others' stories to be heard is indeed the strength of his method. I am one reader who does not need to crowd into his kitchen as well.

From my seat on the bus.[6] Throgmorton's texts come as close as any I know about public decision making to enacting the role of the planner-storyteller and the definition of narrative that best serve the public trust. The storyteller is the one who actively makes space for the story(s) to be heard. An effective story is that narrative which stands the habits of everyday life on their heads so that blood fills those brainy cavities with light. Such a story fully exploits the materials of time (duration, frequency of repetition), time-space (chronotope), and space (scale, perspective, remoteness), deliberately arranging them in unfamiliar ways so that they conscript readers who are willing to suspend their habits of being and come out in the open to engage in dialogue with strangers. To paraphrase de Certeau, they take an equilibrium by surprise.

Whether such storytelling or any storytelling provokes action is a difficult question to answer. I once thought to leave the literature business because I could not answer that question. I now suspect it is better to ask if such a storytelling practice (telling and reading) sustains cities in an environmentally healthy, socioeconomically sound, and morally supportable way. After all, with enough force, some actions can be easy to achieve. The bill is passed; the jail is built. A sustainable urban future is a far more elusive and desirable goal, one that an attentive storytelling practice and analysis can serve.

Editors' Introduction to Chapter 2

When people who are concerned about ecological sustainability revisit Lenin's old question, "What is to be done?" they either advocate politically feasible actions that are too minor to make any difference or else dream of an ecotopian future or deliverance by catastrophe. So observes Tom Athanasiou in Divided Planet *(1996). Throgmorton instead imagines a practical, radical, possible response to the demands of sustainability. Implicit at the core of his argument is art critic John Berger's observation in* The Look of Things *(1974, p. 40, emphasis added) that "it is space not time that hides consequences from us." Drawing on this assertion, which is central to Edward Soja's spatial analysis of the past decades, and on his own scholarship and past experience as an elected official, Throgmorton points to the remote effects of everyone's local habits and advocates radical but practical transformations enabled by our ability to take those effects into account.*

Despite Throgmorton's claim that it is possible and necessary to retain local consciousness of multiple geographic scales, planners employed by representative governments or private firms may well be skeptical about the viability of incorporating multiple scales into their daily practices. The experiences of Franklin and Berkshire, narrated in this volume, raise doubts about the geographic breadth and ecological flexibility permitted planners who serve appointed boards and elected officials representing the citizens of territorially bounded governments. "What can *be done?" they might well ask.*

The will to change . . . has to come from a storyteller's ability to make a narrative and physical space in which to juxtapose multiple, traditional stories so that they enrich, renarrate, and transform that space rather than compete for ultimate control.

—Barbara Eckstein

2

Imagining Sustainable Places

James A. Throgmorton

I can see it clearly in the distance. Although the nuclear power plant is many miles away, its cooling tower glitters in the fire's raging glow. How lucky I am, I think, as I watch the fire's contaminated smoke drift away from me to the right. But then the wind shifts, and bright red cinders begin to waft toward me. Some settle on my arm.

I awake in a pool of cold sweat, my heart racing and my mind still immersed in the dream. Feeling charred, I cannot drift back to sleep. After rising from bed and flicking on the light, I take a cool drink of water. The radiators in my bedroom sit idle, waiting to be activated on the return of winter, while a gentle breeze brings fresh air and the late-spring song of chickadees through my open window. Time to shave and dress. Looking at myself in the mirror, still bleary-eyed, I begin reflecting on the toxic dream. Why does it haunt me so? Perhaps it speaks to my own life and to the place in which I live. Where do the electricity and natural gas that light and warm my room come from? Where do the contaminants produced by their use go, and what effects do they have? Where does that water in my tap come from? Is it safe to drink? Midsummer nights can be brutally hot and humid. Should I comfort myself by installing a window air conditioner?

The dream lingers as I leave my house. Standing on my front steps, I scan the neighborhood. On first glance, it appears to be quite ordinary, a part of middle America where nothing important ever happens, but where many people believe they can live "the American dream" (figure 2.1). However, this morning, influenced by my dream of toxic pollution, I see a complex "organic machine" (White, 1995) mixing "nature" and "culture." Directly across the street stands a modest two-story, wood-frame house. A deep blue sky partly filled with cumulus clouds floats

Figure 2.1
A neighborhood in middle America. (Photograph by James Throgmorton.)

above it, while sewage, water, and cable television lines wend their way below it, just out of sight. (Where do the sewer lines lead? Who and what are affected by the wastes transported through them? Who decides what information will be transmitted on those cable lines? What invisible contaminants permeate the sky?) The street that runs between my house and my neighbor's is paved with old brick and lined with cars on one of its sides. An alley cuts across the street, just to my left. Telephone and electric power lines stretch across and along the street and alley, connecting houses to poles and then disappearing in the distance. Each house has a small green lawn and many trees (one or two of which appear to be rather old). Trash cans and recycling bins stand in front of the houses, awaiting collection. (Where do collectors take this trash, and what becomes of it once it gets there?) Cars occasionally rumble down the street, but far more of them travel along two nearby streets and Interstate 80, which is located a mile and a half north of my house. (Where were those cars manufactured? What social and environmental costs does their production and use entail?) Two young students sit on the porch across the street. We wave hello.

My dream about toxic fallout came to me in June 2000.[1] To a degree, it reflects my personal background and ongoing preoccupations. To that degree, it reveals that I—a middle-class, middle-aged, white American male who has been working on environmental issues since 1971, who teaches urban planning at a research university in a small midwestern city, who has served as a city councilor in my town, and who was in the midst of co-conducting a symposium about storytelling and the sustainability of American cities—am worried that I do not live in a sustainable place. And yet my dream must surely be understood as extending beyond the merely personal. In a deeper sense, it is part of a discourse about sustainability that has engaged people around the world over the past 15 years and more.[2]

So there is a claim embedded in my dream: the way of life that dominates my typically unique American city cannot be sustained. It also implies a hope: that my hometown can become a more sustainable place. The moment I make that claim and express that hope, however, I hear a multitude of voices rising in response, including the following three:

The editors of my hometown newspaper:

The Iowa City Area Development Group [ICAD, a private organization] has a plan [for developing the region's economy]. And it has set goals. And it just might work. . . . We're all familiar with the debates—preserving agricultural land, avoiding sprawl, taking care of traffic, watching for pollution, making sure jobs pay a living wage. We will deal with those questions and more, as we have for years. But there's a bottom line too. We have wonderful services and amenities in our area. They cost more each year. We either grow—with a commercial and industrial tax base—or we stagnate. And then we start to lose some of what makes this a great place to live. (*Iowa City Press-Citizen*, 2001)

Rachel Carson, author of *Silent Spring* (1962):

There was once a town in the heart of America where all life seemed to live in harmony with its surroundings. The town lay in the midst of a checkerboard of prosperous farms, with fields of grain and hillsides of orchards where, in spring, white clouds of bloom drifted above the green fields. . . . Then a strange blight crept over the area and everything began to change. Some evil spell had settled on the community. . . . Everywhere was a shadow of death. . . . There was a strange stillness. The birds, for example—where had they gone? . . . It was a spring without voices. . . . In the gutters under the eaves and between the shingles of the roofs, a white granular powder still showed a few patches; some weeks before it had fallen like snow upon the roofs and the lawns, the fields and streams. No witchcraft, no enemy action had silenced the rebirth of new life in this stricken world. The people had done it themselves. (1962, pp. 1–3)

LeAlan Jones, an African-American youth growing up in a public housing project on the South Side of Chicago: Pondering what might have caused two young boys to throw a 5-year-old child off the fourteenth floor of a building in the projects, he says:

Those boys didn't value life. Those boys didn't have too much reason *to* value life. . . . What else do you think would happen when they grow up in a concrete world. . . . A kid deserves something better than that. . . . As children [who grow up in the projects], they have to make day-to-day decisions about whether to go to school or whether to go on the corner and sell drugs. As children, they know that there may not be a tomorrow. Why are African-American children faced with this dilemma at such an early age? Why must they look down the road to a future that they might never see? . . . I know you don't want to hear about the kids getting shot in "that" part of the city. But little do you know that "that" part of the city is your part of the city too. This is our neighborhood, this is our city, and this is our America. (Jones and Newman, 1997, pp. 141, 145, 199, 200)[3]

The editors of my hometown newspaper claim that "our area" is "a great place to live"; Rachel Carson tells a fable in which "a strange blight," imposed by the people upon themselves, is silencing "the rebirth of new life" in "a town in the heart of America"; and LeAlan Jones claims that other young men in the Chicago projects (which is "your part of the city too") don't have much "reason to value life" largely because they must "look down the road to a future that they might never see." Juxtaposed against one another, these responses begin to reveal the complexity of envisioning and creating places that can be sustained, and they raise four intertwined questions that anyone who promotes sustainable places must grapple with:

1. What should be reborn tomorrow?
2. What are the boundaries of "our place"?
3. How can we build a future that gives us a "reason to value life"?
4. Who are "we"?

What Should be Reborn Tomorrow?

By expressing a deep personal and cultural fear about technological threats to ecological sustainability, my dream about the nuclear power plant is in part a residue of 1970s environmentalist ways of thinking. Back then, environmentalists argued that continued exponential growth in a finite environment (building more nuclear power plants to meet a 7 percent increase in demand for electricity per year, for example) would soon encounter natural limits. Too much pollution would be generated, nonrenewable resources would be depleted, and/or the biodiversity found in the few remaining natural ecosystems would be dramatically reduced. Unless population and economic growth were constrained, they believed, natural environmental limits would be exceeded, major ecosystems would collapse, and both humans and nonhumans would experience great harm. Inspired in large part by Rachel Carson's *Silent*

Spring, this tragic limits-to-growth narrative claimed that all of us living on the planet Earth were looking down a road we might never see.

Sustainable Places

As the 1980s progressed, however, many environmentalists concluded that the limits to growth argument was losing credibility, partly because global economic trade seemed to eradicate local limits to growth, but more important because it seemed to dismiss claims that economic growth was needed to alleviate starvation, disease, and poverty, especially in lesser developed countries of the global South and in places like Jones and Newman's part of the city. These environmentalists began working with others to reformulate their argument into a "win-win" scenario that would transform the tragedy of the commons into a romance. This effort culminated in 1987 with publication of a report by the World Commission on Environment and Development (the Brundtland Commission), *Our Common Future* (1987). This report advocated "sustainable development," which was defined as "development that meets the needs of the present without compromising the ability of future generations to meet their own needs" (1987, p. 43). It would be "a type of development that integrates production with resource conservation and enhancement, and that links both to the provision for all of an adequate livelihood base and equitable access to resources" (1987, pp. 39–40).

Five years later, at the United Nations Conference on Environment and Development (or Earth Summit) in Rio de Janeiro, most heads of national governments included the idea of sustainable development in a package of agreements, including a biodiversity convention, a climate change convention, a statement on forest principles, an agreement to work toward a desertification convention, the Rio declaration on environment and development, and Agenda 21, an 800-page plan for saving the planet.

Motivated partly by these two events, but more important by the fear that conventional patterns of economic growth cannot be sustained and the hope that fundamental changes might be accomplished, a wide variety of organizations in Europe, North America, Australia, and elsewhere (e.g., Sustainable Seattle, Sustainable Bay Area, and the European Sustainable Cities and Towns Network) have sought to make their places more sustainable (President's Council, 1996; Urban Ecology, 1996; Beatley and Manning, 1997; Charter of European Cities, 1994; Sachs et al.,

1998). As an elected member of my hometown's city council, I made sustainability a central element of my political practice. However, one cannot make a place more sustainable without having some sense of what "place" means, and it turns out that the idea of place has multiple dimensions. As the literary critic Lawrence Buell (2001, p. 62) puts it, "[a] place may seem quite simple until you start noticing things."

The idea of place has a long and complicated history. Drawing upon that history, Buell suggests that there are five dimensions of place connectedness. The first is concentric areas of affiliation (home, neighborhood, town), decreasing in intimacy as one fans out from a central point. The editors of my hometown newspaper seem to have had this dimension in mind when they claimed that "this" (presumably the Iowa City area) is "a great place to live," and one can easily imagine a "sustainable place" initiative starting at this central point. Indeed, it almost must, for it is at this level that people seem most willing and able to express their strongest sense of attachment to and care for places. Guided by this sense of caring attachment, people try to maintain the quality of their homes, the historic character of their neighborhoods, the vitality of their social communities, and the viability of farmland and locally prominent natural ecosystems from a range of threats.

However, operating exclusively in this dimension of place can, and often does, generate worrisome consequences. If the area of affiliation is limited to the official boundaries of one's city, it can easily lead to a fear of contamination, the construction of "gated walls" around the city, the exclusion of unwanted neighbors (e.g., landfills; low-wage earning families with school-aged children; people of the "wrong" race, class, or ethnicity), and the export of unwanted wastes (e.g., air and water pollution, criminals). It can also lead local elected officials and their technical staff to focus on maximizing the self-interest of their territorially defined city in the context of a highly competitive global economy and local budgets dependent on property taxes.

A second dimension of place as Buell (2001) sees it would be a scattergram or archipelago of locales, some perhaps quite remote from one another. As he puts it, "to understand fully what it means to inhabit place is therefore not only to bear in mind the (dis)connections between one's primary places but also the tenticular radiations from each one" (2001, p. 66). When I scan my neighborhood and see electric power lines, sewer lines, trash cans, and cars on the street, and when I reflect on my hometown editors' desire to expand the tax base by recruiting new business investment, I am beginning to recognize the "tenticular radiations" from my hometown. Places are embedded in complex technosystems and environmental pathways that tie distant places to one another, poke holes in the walls between cities (Frug, 1999), and disrupt the radical polarity between cities and pristine nature (Cronon, 1996) (figure 2.2 a–c). Rachel Carson seems to be pointing to this dimension when she refers to the strange blight that crept over the area, as does LeAlan Jones when he claims that his part of the city is your part of the city too.

The third dimension Buell points to is that places are continually shaped and reshaped by forces from both inside and outside; they have histories and are constantly changing. For those who have developed a sense of place, these changes superimpose upon the visible surface "an unseen layer of usage, memory, and significance—an invisible landscape, if you will—of imaginative landmarks" (Buell, 2001, p. 67, citing ecocritic Kent Ryden, 1993).[4] LeAlan Jones understands this dimension when he and his friend Lloyd stand at the foot of the fourteen-story building where the 5-year-old boy who had been pushed met his death: "We're standing right here where the little dude got killed," Jones says. "How's that make you feel? That's like walking over a graveyard, ain't it?" Lloyd replies, "LeAlan, why does it make you feel different standing right here? You ain't standing on his grave or nothing, you're just standing on grass. It happened right here, but it ain't right here." Jones responds, "That's like Indian burial ground, homie—you ain't gonna catch no Indian walking on another Indian's grave or where he died. It's just a feeling, man. You never know—Shorty's soul could be looking down at us now, doing a story about him" (Jones and Newman, 1997, p. 103).

Buell's fourth dimension derives from the fact that people are constantly moving into or out of places. Thus any one place contains its residents' accumulated or composite memories of all the places that have been significant to them over time. Drawing heavily on her knowledge of Detroit, for example, June Manning Thomas (1994) argues that one cannot comprehensively understand the history of American cities and their planning without understanding the black urban experience.

That experience began with hundreds of thousands of southern black workers migrating northward between world wars I and II, only to be met by racially restrictive zoning ordinances and covenants, and by white

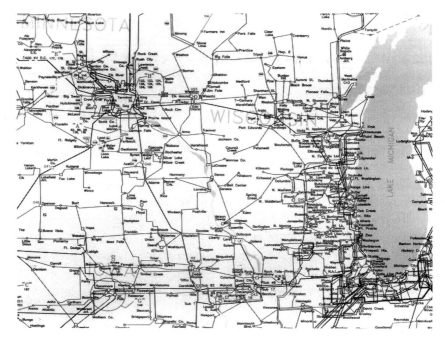

Figure 2.2
Tenticular radiations connecting one midwestern American neighborhood to distant places. (a) The upper midwest region's electric power transmission system. (Source: Mid-Continent Area Power Pool, 2001; reprinted with permission of Midwest Independent Transmission System Operator, Inc.)

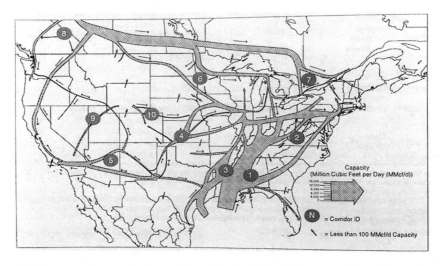

(b) The natural gas supply corridors for the United States and Canada, 1997. (Source: U.S. Energy Information Administration, 1998.)

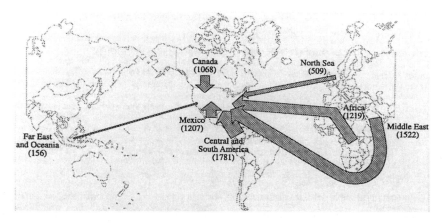

(c) The worldwide flow of crude oil into the United States (thousands of barrels/day, 1996). (Source: U.S. Energy Information Administration, 1997.)

riots against blacks. In the 1950s and 1960s, public housing, urban renewal, and interstate highway projects were sited in such a way as to confine blacks within existing ghettos or displace them from land that local political regimes wanted for other uses. Confined to ghettos but disaffected by decades of racially insensitive policies, black neighborhoods exploded in violence during the 1960s' decade of civil rebellion. And now blacks find themselves in a racially divided metropolis, living in cities full of despair and oppression. Though nearly inundated by the tide of urban decline, Thomas concludes, many black elected officials and communities carry out heroic efforts to preserve and improve their neighborhoods.

Finally, Buell suggests that fictive or virtual places can also matter. Kim Stanley Robinson's novel *Pacific Edge* (1990) and his trilogy of novels about Mars (1993, 1994, 1996) exemplify how fictive places can fit into the discourse about sustainable places. *Pacific Edge* imagines an environmentalist-utopian reconstruction of southern California around 2065. Political conflict and compromise continue, as does technological innovation, but they do so in an altered, ecotopian, context. In *Pacific Edge*, inhabitants discover a way to live in a place sustainably.

Robinson's later Mars trilogy imagines a 150-year future history about "terraforming" the red planet. In the Mars trilogy, settlers at first continue the American myth of "the West," abandoning their past (Earth and its conflicts) and pushing the frontier onto a distant planet. However, as their future unfolds, they discover that it is impossible to leave the past behind, that treating Mars as an escape valve for Terrans effectively undermines the ability to sustain good lives on Earth, and that being on Mars transforms them; it "areoforms" them. They learn that for life to thrive on both Earth and Mars, they have to negotiate differences, both among themselves and with Terrans.

One can also sense a certain fictive quality to contemporary discourse about actual places. When advocates of ecological sustainability speak of Curitiba, Brazil; Gaviotas, Columbia; Freiburg, Germany; and Portland, Oregon, they emphasize how roughly 20 years ago the citizens of those places explicitly chose to pursue innovative development paths. Now when advocates speak of those places, they imagine that they have come very close to the ideal of sustainability. Writing of Curitiba, for example, Hawken et al. say, "its results show how to combine a healthy ecosphere, a vibrant and just economy, and a society that nurtures humanity. Whatever exists is possible; Curitiba exists; therefore it is possible. The existence of Curitiba holds out the promise that it will be the first of a string of cities that redefine the nature of urban life" (1999, p. 308). It is almost as if Robinson's environmentalist-utopian vision has come to life, as if his future history has been condensed in time.

These five dimensions of place combine. Thus, Buell argues, "If . . . one thinks of place sense as containing within it many different patches besides just home, including what comes to us via the world of images as well as through live transactions, plus the changes in us relative to place and the awareness of landscape as timescape, then we are on the way to arriving at a conception more fit for local, regional, and global citizenship" (2001, p. 77).

What Are the Boundaries of "Our Place"?

All five dimensions of place are relevant to creating sustainable places. For present purposes, however, I want to focus on the second, tenticular radiations, partly because it is the dimension I found most difficult to act upon effectively as a city councilor, but primarily because it gets to the heart of the questions raised earlier, especially the one that asks, "What are the boundaries of 'our place'?"

Consider Tenticular Radiations

Much like my hometown editors, most business people and elected officials (in the United States at least) believe that economic growth is the only way to generate enough wealth to lift people out of poverty and to provide environmental amenities. Acting in a context of a globalized economy, they believe they must do whatever is necessary to attract and retain investment. They further believe that people freely express their preferences for particular goods and services and that prices give them all the signals they need to make choices. For them, a sustainable place is one that experiences "sustained growth."[5]

The claim that sustained growth equals sustainability (that "we either grow or stagnate") ignores the fact that negative externalities (e.g., air pollution, traffic congestion, and other adverse effects on third parties) are not included in the prices of goods and services. Consequently, consumers consume more resources and produce more pollution than economists would consider optimal. Recognizing the pervasive presence of such negative externalities in the market, many eco-economists argue that most environmental problems will disappear if we get the prices right; that is, if we include the cost of externalities in the prices of goods and services (i.e., internalize them).

For people who seek to create sustainable places by acting at the local level, however, the problem is that localities cannot get the prices right without placing themselves in an uncompetitive market position relative to neighboring municipalities or states, at least in the short run. If "green taxes" are adopted at the city or county level, consumers will simply cross boundaries and purchase goods and services in a nearby city or county. However, unless negative externalities are internalized in the prices of goods and services, consumers cannot take into account the unpriced consequences of their choices. Worse perhaps, they might not even know that their choices are producing adverse effects. As the philosopher of technology Aidan Davison (2001) puts it, technological systems move commodities (and their prices) to the foreground of our experience, while causing the technosystems themselves (and their adverse effects) to recede from view.

Spatially and psychologically disconnected from the resources that sustain them, consumers and citizens drift blissfully along in a kind of sleepwalking that Langdon Winner (1986) calls "technological somnambulism," and they display a tendency to repress or exaggerate environmental effects and connections. In Buell's (2001) view, this tendency constitutes a negative manifestation he calls an "environmental unconscious"; that is, the impossibility of becoming fully conscious of environmental connections.[6] The combination of remoteness (Plumwood, 1998), technological sonambulism, and an environmental unconscious hinders our ability to see what truly matters in our lives and inhibits our ability to see that there are other possible ways of living (Davison, 2001). Considered more positively, however, the combination also offers the possibility of change; it contains "a residual capacity . . . to awake to fuller apprehension of physical environment and one's interdependence with it" (Buell, 2001, p. 22).[7]

How can people act wisely in the here and now, given the pervasive presence of unpriced costs, the capacity of technosystems and complex environmental pathways to produce remote effects, and the capacity of technological somnambulism and environmental unconsciousness to reduce our awareness of these effects? How can tenticular radiations from a place (and their remote ecological and social effects) be brought out of the environmental unconscious and into locally based discourses about sustainability? How can we awaken from our toxic dream?

How Can We Build a Future that Gives Us a Reason To Value Life?

To counter the diluting effects of environmental unconsciousness, Buell suggests that we need to engage in "acts of environmental imagination." I want to draw attention to three such acts, all of which could be thought of as tropes in a potentially persuasive story (Throgmorton, 1996) about the future of American cities: to leave modest ecological footprints, construct a factor-10 economy, and build the Regional City.

Leave Modest Ecological Footprints
An ecological footprint (Wackernagel and Rees, 1996) is a rather simple heuristic device that indicates the land area necessary to sustain current levels of resource consumption and waste discharge by any given population. To facilitate understanding of the concept, Wackernagel and Rees ask their readers to imagine an experimental city:

First, imagine what would happen to any modern city or urban region—Vancouver, Philadelphia or London—as defined by its political boundaries, the area of built-up land, or the concentration of socioeconomic activities, if it were enclosed in a glass or plastic hemisphere that let in light but prevented material things of any kind from entering or leaving. . . . It is obvious to most people that such a city would cease to function and its inhabitants would perish within a few days. The population and the economy contained by the capsule would have been cut off from vital resources and essential waste sinks. . . . [Alternatively] let's assume that our experimental city is surrounded by a diverse landscape in which cropland and pasture, forests and watersheds—all the different ecologically productive land-types—are represented in proportion to their actual abundance on the Earth, and that adequate fossil energy is available to support current levels of consumption using prevailing technology. Let's also assume our imaginary glass enclosure is elastically expandable. The question now becomes: how large would the hemisphere have to become before the city at its center could sustain itself indefinitely and exclusively on the land and water ecosystems and the energy resources contained within the capsule? (1996, pp. 9–10)

According to Wackernagel and Rees, the size of a city's ecological footprint depends on its population and its per capita rates of consumption and waste production (which in turn vary with income, prices, social values, and technological sophistication). They estimate that the average U.S. American produced an ecological footprint of 5.1 hectares in 1991, an area far larger than foortprints produced by people in other parts of the world. Implicitly referring to tenticular radiations, they say, "*the ecological locations of human settlements no longer coincide with their geographic locations.* Modern cities and industrial regions are dependent for survival and growth on a vast and increasingly global hinterland of ecologically productive landscapes" (1996, p. 29). "[A]s living standards rise, more and more people live on ecological carrying capacity 'imported' from somewhere else. The obvious follow-up question is: how long will it be before we run out of 'somewhere else?'" (1996, pp. 155–156).

Public discussion of a place's ecological footprint and the possibility of "us" running out of "somewhere else" could well be a useful starting point in expanding local understandings of the meaning of "sustainable place." Indeed, while guiding a study abroad course in Europe in 1998, I learned that many European cities actively consider their footprints when deliberating local actions (see also Beatley, 2000). However, as a former city official, I am confident that footprints by themselves would not carry much weight in local decision making, at least not in the United

States. Footprints would have to be combined with other acts of environmental imagination.

Construct a Factor-10 Economy

Recent arguments for ecological modernization, industrial ecology, dematerialization, decarbonization, and a factor-10 economy would bolster calculations of ecological footprints. Hawken et al. in *Natural Capitalism* (1999) present these arguments in a clear and compelling fashion. Building on an ecological engineering understanding of efficiency rather than that of environmental economists, they argue that their approach would enable wealthy societies, such as the one in which my neighborhood is embedded, to reduce their per capita consumption of resources and production of pollution by 90 to 95 percent. Creating this more efficient "factor-10 economy" would open up "environmental space" (Sachs et al., 1998) for less wealthy societies to increase their per capita wealth while decreasing their rate of population growth and preserving biodiversity.[8]

Hawken et al. claim that wasted materials, labor, and wealth are interlocked symptoms of one problem: using too many resources to make too few people more productive. "[I]t is helpful to recognize that the present industrial system is," they say, "practically speaking, a couch potato: It eats too much junk food and gets insufficient exercise. In its late maturity, industrial society runs on life-support systems that require enormous heat and pressure, are petrochemically dependent and materials-intensive, and require large flows of toxic and hazardous chemicals. These industrial 'empty calories' end up as pollution, acid rain, and greenhouse gases, harming environmental, social, and financial systems. . . . It has been estimated that only 6 percent of [the U.S. economy's] vast flows of materials actually end up in products" (1999, pp. 14–15). In their view, this waste needs to be seen not as a problem but as an opportunity; it can be greatly reduced without compromising our well-being. "Any improvement that provides the same or a better stream of *services* from a smaller flow of *stuff*," they say, "can produce the same material wealth with less effort, transportation, waste, and cost" (1999, p. 62).

In their view, most of this striking 90 to 95 percent reduction in use of materials and energy can come from systematically combining a series of successive savings. Of particular interest is the way they implicitly link these reductions to tenticular radiations from home places and to the ecological footprints that home places impose: "Often the savings come in different parts of the *value chain* that stretches from the extraction of a raw resource, through every intermediate step of processing and transportation, to the final delivery of the service (and even beyond to the ultimate recovery of leftover energy and materials). The secret to achieving large savings in such a chain of successive steps is to multiply the savings together, capturing the magic of compounding arithmetic" (1999, p. 177, emphasis added).

On first reading, their claim seems quite implausible. However, it becomes less so the more one connects it to daily life. "See that car across the street," they might say to me as I scan my neighborhood from my front porch. "Did you know that the average American car is parked about 96 percent of the time? When used, moreover, that car is embarrassingly inefficient. Of the energy in the fuel it consumes, at least 80 percent is lost, mainly in the engine's heat and exhaust, so that at most only 20 percent is actually used to turn the wheels. Of the resulting force, 95 percent moves the car, while only 5 percent moves the driver." By making cars ultralight, ultra-low drag, and powered by an electric hybrid system, we could reduce, at a profit, close to two-thirds of America's carbon dioxide emissions while preserving the mobility, safety, performance, and comfort of traditional cars.

Put that way, the argument by Hawkins et al. becomes more persuasive. It reveals a means by which a place's ecological footprint could be dramatically reduced without harming its economy. Nevertheless, it too is not sufficient, partly because economic growth and technological change might soon overwhelm improvements in efficiency, but also because home places are tightly intertwined with national and transnational firms, elected officials, and others who will contest the merits of improved efficiency and resist major moves toward it. More important, perhaps, the factor-10 approach would do very little to alter land development patterns that force people to drive motor vehicles, require construction of ever-more costly infrastructure, and divide regions into hypersegregated enclaves of rich and poor, black and white, opportunity and despair. Meritorious insofar as it goes, the factor-10 economy would not alleviate unjust social conditions that give hundreds of millions of people in the world little reason to value life. As long as that is the case, no place can truly be considered sustainable.

Build the Regional City

To a degree, the sustainability of places is related to the spatial patterns in which they develop: sprawling low-density development can overwhelm short-term improvements in use of materials; and spatial inequities in wealth, opportunity, and risks can undermine a region's ability to thrive over time in a globalized society. Thus New Urbanist proposals for changing the physical design of American neighborhoods, towns, and cities constitute a third act of environmental imagination that might lead to sustainable places.[9]

A particularly important feature of the New Urbanists' argument is their effort to place neighborhoods and towns in the context of "the region," and to place the region in the context of economic globalization and global environmental systems. Peter Calthorpe and William Fulton's (2001) *The Regional City* displays this effort most clearly. As they put it,

Since the end of the Cold War, as the "globalization" of our economy has accelerated, the metropolitan region has come to be viewed as the basic building block of this new economic order. In today's global economy, it is regions, not nations, that vie for economic domination throughout the world. In addition, our understanding of ecology has matured rapidly, as we have come to realize that the region is also the basic unit in environmental terms. Because of the interconnected nature of ecosystems, we are hooked together with our neighboring communities whether we like it or not. Finally—and perhaps most important from our point of view—we are beginning to set aside our outdated view of independent towns and suburbs and coming to see that the region is also a cohesive social unit. . . . Old or young, rich or poor, the people of every metropolitan region are bound together in ways that greatly affect their daily lives. (2001, pp. 16–17)

As these words imply, Calthorpe and Fulton do not think of suburban sprawl and urban decay as being separate and unrelated problems. Instead, they would argue, a great place to live on the exurban periphery of a large city like Chicago is intimately connected with the concrete world in which people like LeAlan Jones "make day-to-day decisions about whether to go to school or . . . sell drugs" (Jones and Newman, 1997, p. 199). Conceiving of the metropolitan region as a series of interconnected places, Calthorpe and Fulton argue that we need to imagine, design, and construct the "Regional City."

The design of the New Urbanists' Regional City would be based on a few key principles. Several of them focus on physical elements: high-quality natural areas would be identified and preserved; growth boundaries would limit the spatial expansion of development for the foreseeable future; a high-quality public transportation system would provide a spine around which land development would be oriented; and "communities of place" that are walkable and human scaled would be constructed around the public transportation system's major stations.[10] These communities of place would be diverse in population and would contain mixed-use centers and memorable public spaces. Grayfields (such as old, abandoned shopping malls in existing suburbs), greenfields (new developments on the exurban periphery), and brownfields (abandoned industrial sites in older urban areas) would all be redeveloped as communities of place in a regional context. However, construction of the Regional City would also require addressing questions of social and economic equity. "The Regional City will not overcome inequity," Calthorpe and Fulton say, "unless its leaders pursue a regional strategy of deconcentrating poverty, providing adequate affordable housing in proximity to jobs, and creating a more equitable distribution of investment throughout the entire metropolitan area" (2001, pp. 73–74). In their view, such a strategy would include regional fair-share housing, regional tax-base sharing, and urban educational reform.

"Look at the neighborhood around you," Calthorpe and Fulton might say to me if we were traveling together toward my routine places of work. "Consider how its physical design structures the ways in which you can travel and who you might encounter while you do so. If you had chosen to live in the exurban periphery of any typical American city, conventional patterns of development (including the lack of public transit and entrenched fears of race- and class-related conflicts) would force you to drive and make it virtually impossible to encounter people who are significantly different from yourself. Since you live quite close to your place of work, however, and since you can use diverse modes of transportation, you are far more likely to encounter diverse others as part of your daily life. Even so, your neighborhood is embedded in a region that is almost completely dependent on motor vehicles, could scarcely be described as racially or ethnically diverse, and is divided into local government units that behave as if they were autonomous individuals rather than part of an interconnected region. Creating a sustainable place requires more than altering your own behavior or using an ultra-efficient electric-hybrid car; it also means transforming your region's built environment to support nonautomotive modes of transport and diversified populations within neighborhoods. A sustainable place enables all residents to look down the street and see a future worth living."

For the purposes of this chapter, the important feature to note about Calthorpe and Fulton's argument is that it invites people to think of themselves as living in a region, not in isolated neighborhoods or towns. In effect, it asks the editors of my hometown newspaper, adherents of Rachel Carson's work, and people like LeAlan Jones to imagine themselves as a spatially expansive "we" that can jointly create "our place" and thereby give all of us a reason to value life. This is not a simple request, and not all people respond to it positively. A recent letter to the editor of the *Louisville Courier-Journal* illustrates how negative the response can be. Criticizing Kentucky Governor Patton's proposed "smart growth policy," Paul E. Sloan, Sr. (2001) opines,

If the American people haven't seen how far the Democratic Party has turned to the left politically, this smart-growth policy should be a sure sign. This policy, if implemented in Kentucky, will eventually tell me that I can't live in the suburbs on one to five acres on a golf course, but have to live where the state government has assigned in its smart-growth policy. . . . We don't need this Governor telling landowners in Kentucky to whom they must sell their properties, or be penalized. President Ronald Reagan defeated communism, and we don't want it creeping back in this country through these Democratic liberals.

Although I find Calthorpe and Fulton's argument to be timely and deeply compelling, it would appear from Mr. Sloane's letter and many other statements like it that many—perhaps most—Americans would not. It would seem that most are not currently ready to turn away from the automobile, buy homes in socially and economically diverse neighborhoods, and thereby construct the Regional City. My sense is that we Americans will not collectively be willing or able to make these changes as long as most of us remain unconscious of (and do not take into account) the remote effects of our actions.

What would make us sufficiently conscious of those remote effects? War, revolution, or some unpredictable natural disaster that initiated a rapid and large price shock is surely one possibility. Lacking that, what might work? I would suggest we need to hear stories that might persuade us that such a change is desirable. I would further suggest that such persuasive stories are most likely to be told and heard during quotidian encounters, and that such encounters are most likely to occur in public places. If we have the opportunity to walk within places, to encounter people who differ from ourselves, and to hear diverse stories of everyday practice, we will be far more likely to learn that we live in different spaces and that our local actions produce remote effects that should be taken into account.

Conversely, if we do not encounter one another routinely as part of our everyday lives, we will grow increasingly ignorant, fearful, and distrustful of one another. When struck by deeply felt emotions, such as fear, anger, loss, grief, and greed, we will have no (or at least a highly atrophied) informal public means for understanding why we have such different views for processing our emotional reactions collectively, or for resolving the conflicts that come with them. We will find ourselves crying and shouting at one another in more formal settings, completely unable to understand our differing points of view.

My worry about the New Urbanists' Regional City is, therefore, that its creation presumes public places and inclusive deliberative processes that enable people to encounter diverse stories as a part of ordinary life, but that people are not likely to encounter such stories unless the Regional City's public spaces and inclusive processes already exist. This worry becomes even deeper for stories that circulate in remote places where the region's tenticular radiations impose costs that go unpriced and unnoticed, at least back in the region. Without inclusive places and processes, most of us will continually sleepdrive through familiar locales, seeing them through familiar thoughts, forever trapped in a "walled" landscape of banal repetition, unconscious of the adverse effects we impose, and never able to find what we do not know we are looking for (Solnit, 2000). Most of us will continue to segregate land uses in space, build gated communities, exclude unwanted neighbors, build a privatized city (Lofland, 1998), and live in sprawling developments that impose massive footprints on distant places, times, and people. We might think we are living in a great place, but it would be a place scarcely worth sustaining.

Who Are We?

The line of reasoning issuing from my dream about toxic fallout seems to lead to an impasse. The technosystems and complex environmental pathways of the "organic machine" in which we are embedded combine to produce remote effects, while unpriced negative externalities, technological sonambulism, and environmental unconsciousness diminish our awareness of them. To construct sustainable places, diverse people must be able to encounter one another as part of their everyday practice, hear one another's stories, and hence become conscious of those remote effects. However, the public spaces required for such encounters to occur will not exist unless we build something like the Regional City. How, then, can we create sustainable places when so many people in America's metropolitan regions live behind "walls," scarcely know one another, transfer social and ecological costs to distant places and times, consume environmental space needed for development in the global South, and leave millions of people with little reason to value life—all the while unconsciously fearing that those costs will return to contaminate them? To answer these questions well, we have to think clearly about who we are and who we want to be.

The editors of my hometown newspaper say, "We will deal with those questions . . . as *we* have for years" (emphasis added). Politically, that is precisely what advocates of sustainable places fear: that undemocratic institutions will continue to pursue sustained growth as they have for years.[11] A different form of politics and different understandings of who "we" are and what "our place" is are necessary in order to bring acts of environmental imagination such as ecological footprints, the factor-10 economy, and the Regional City into deliberative processes at every level of public action.[12] To build sustainable places, we need to create discursive processes that make space for stories that "defamiliarize the familiar" (Eckstein, chapter 1, this volume) and thereby expand our sense of who we are, what our place is, and what gives us a reason to value life.

The first thing to understand is that we define the region, while the region defines who we are. In other words, constructing a persuasive story about the region is also a matter of constructing regional identity; the "we" the story constructs depends on how the story is spatialized (Soja, chapter 10 in this volume). As legal scholar Gerald Frug (1999) documents so clearly, the current structure of governance in America constructs the region as an aggregation of autonomous local governments. This structure encourages local governments to compete with one another to attract businesses and wealthy households, to rely heavily on exclusionary zoning, to emphasize property rights, to structure tax incentives to benefit individual local governments, and to transfer hidden and uncounted costs to distant places and peoples. If the "we" implied by the story is limited to people who live in these local government units—to well-off residents (like Mr. Sloane perhaps) who live in gated exurban communities—the story is likely to be an exclusionary one. Alternatively, if the "we" implied by the story is expanded to include the residents and users of a much larger area, as well as distant people affected by actions that take place in the region, then the story is likely to be far more inclusive, and the members of the region will be far more likely to take their effects upon one another into account; they will be more capable of internalizing more of their negative externalities.

The crucial first step is, therefore, for some people to imagine a region whose boundaries are socially and spatially more inclusive than current ones. The next step would be to organize a forum in which diverse people of the region could begin talking with one another. In many cases, such as in my home region, where ICAD seeks to attract business to the "Iowa City *area*" (emphasis added), such forums already exist but have a very narrow constituency. If possible, these existing forums should be diversified. If not, new ones should be created.

Unless there is pressure from below, from constituents, it is unlikely that a sufficient number of politically influential people will be willing to participate in a new regional forum. Frug (1999) suggests that the organizers of the new regional forum—those who imagine the region—should provide that pressure by building on the fact that no local government unit consists of a unified "we." He encourages inventors of the region to deconstruct conventional claims that there is a unified "we" within cities—that Mr. Sloane speaks for everyone—and to construct political alliances around pressing practical problems that cut across existing governmental boundaries. Traffic congestion, water pollution, economic vitality, and affordable housing are but a few examples.

Once a new and inclusive regional forum has been created, some members of the region will have an opportunity to present and justify their claim that a more sustainable region needs to be built. They will have an opportunity to transform understandings by conveying a sense that (1) we are part of an interconnected region; (2) we have been unconsciously imposing an ecological footprint that consumes the environmental space of people who live in developing parts of the global South; (3) we have a moral responsibility to reduce that footprint by dramatically increasing the efficiency with which we use materials and energy; and (4) we should create a more compact, diverse, and public transit-oriented regional development pattern. They will be able to say that taking these actions will enable the region to achieve enduring efficiency, thrive in a context of economic globalization, and provide an enduring opportunity for all of us (including distant people who are affected by our actions) to live good lives.

As these storytellers advocate their vision of a sustainable region, they will have to be continually aware that they are both creating a new structure and being affected by the existing structure (Bourdieu, 2000; Throgmorton, 2000); as they "terraform," they are being "areo-formed." They must begin by articulating a coherent vision, but they will then have to practice that vision within webs of relationships and the material realities of particular places at particular moments in time. They will have to wind their way through the "muck" of zoning codes, comprehensive plans, budgets, capital improvement programs, tax incentives, national transportation subsidies, "super-majority" requirements, election campaigns, pervasive dreams and fears, and all the other actually existing features of contemporary American society and politics (Throgmorton, 2000). They will have a long road to walk, but walking it will be healthier and more joyful than sleepdriving into "a spring without voices."

In the end, therefore, my dream tells me that to care skillfully for our shared world in a context of tenticular radiations, unpriced costs, remoteness, and an environmental unconscious, we have to make space for stories that draw attention to the region's ecological footprint, to increasing the efficiency of the value chains that produce that footprint, and to developing a shared sense of moral purpose in a Regional City. Making space for such stories will require both inclusive spaces (processes) and inclusive places, for in the end the content of the story (and the planning that contributes to it) depends on who authors it. Lacking such stories, people might live in a great place, but it would be a place that stands in the "shadow of death" and gives kids like LeAlan Jones and Lloyd Newman little reason to value life. They, or I should say we, deserve something better than that.

Editors' Introduction to Chapter 3

Beauregard carefully develops a prescription for a sustainable city: it must be a democratic city, that is, one in which diverse, active citizens engage in public storytelling. As a goal, urban sustainability should be understood as an antidote to indifference on the one hand and sweeping self-destruction (urban renewal, sprawl) on the other. In order for discursive democracy to do the work of approaching this goal, private stories must become public.

This prescription does not include a full protocol for interpreting diverse, public stories. Rather, it implicitly returns the reader to Eckstein's analysis, for example, her discussion of Alessandro Portelli's interrogation of trust and truth in oral history work (1991). It also casts a line out to the discursive diversity presented in part II, which is most evident in the work of Rotella, Reardon, Berkshire, and Barthel.

Similarly, Beauregard leaves unanswered questions about the role of government in the sustainable city. Gerald Frug's suggestions for regional governmental structures, in City Making *(1999), complement Beauregard's vision, as does Leonie Sandercock's invitation to a therapeutic planning practice in chapter 6 in this volume.*

How can we awaken from our toxic dream?

—James Throgmorton

3

Democracy, Storytelling, and the Sustainable City

Robert A. Beauregard

Throughout history, scholars have cast the city as the site of a robust, even if at times threatening, democracy.[1] Examples range from the Greek agora where citizens debated common concerns to Thomas Jefferson's warnings about the tendency for cities to embolden mobs to the current academic fascination with flaneurs, public spaces, and urban citizenship. In all cases the issue is the nature and desirability of an active public.

A democracy requires engaged citizens. Open and frequent elections, a vigilant media, and knowledgeable citizens stand in opposition to bureaucratic indifference and the tyranny of minorities. An active public also plays a central role in addressing public ills. Widespread deliberations connect civil society and the state at multiple points and provide the state with the support and legitimacy it needs to act in times of crisis and address societal problems. Only with such broad legitimacy can the state resist the forces of injustice and inequality (Bellah et al. 1991; Putnam, 2000).

In such deliberations, public storytelling is essential. Storytelling enables people of all backgrounds and abilities to frame a sense of what is, reflect on what needs to be done, and then engage with others about the sensibility of their stories. "Stories organize knowledge around [both] our need to act and our moral concerns" (Marris, 1997, p. 53). A democracy without such storytelling is one in which technocrats rule and bureaucracy is unaccountable to citizens. Political leaders lose touch with their constituents and citizens are turned into either clients, passive voters, or both.

The purpose of this chapter is to reflect on democracy and storytelling in the context of a sustainable city, a city that renews its resources and capacities for present and future generations. In the absence of a robust democracy, economic growth and prosperity are likely to become parasitic, degrading the environment and retarding social equity. Sustainability is one of the victims. Elites might live well, but many others will bear the costs of the inattention to shared needs. A sustainable city must be a democratic city.

Discursive Democracy

Back in the mythic 1960s, the city witnessed a participatory democracy in which residents joined together to wrest power from government and resist corporate indifference and greed. The city was viewed as a site of struggle against poverty, racism, and institutional arrogance. The goal was to gain power, and participation meant being there: in the streets, on the steps of city hall; and in welfare offices in order to complain, disrupt, protest, and demand (Chafe, 1986; Gale, 1996).

More recently, participatory democracy as an ideal has been joined by discursive democracy (Bohman, 1996; Calhoun, 1993; Habermas, 1996b; Young, 1996). Instead of offering their bodies for the cause, residents of the city offer their voices. They speak out about their marginalization, about daily injustices, and about the failure of institutions to acknowledge their identities and give them respect. The city is no longer solely a site for contesting power. The struggles are different; they are less about control over institutions and support of interests than about recognition. Earlier concerns with poverty, slums, and institutional discrimination are less prominent than movements organized around issues of identity and rights (Castells, 1997; Fraser, 1995; Garber, 2000; Holston, 1999; Katznelson, 1996; Schudson, 1998).

Compare participatory and discursive democracy with the representative democracy out of which the modern planned city was conceived in the late nineteenth and early twentieth centuries. Street protests around planning issues were few and citizen involvement in policy making mainly involved enabling economic and social elites to influence political leaders. A representative democracy combined institutional capacity and singularity of purpose to enable large-scale projects to go forward relatively unhindered.

By contrast, a discursive democracy threatens the prerogatives of elected officials, economic and cultural elites, and technocrats. The views of these groups are countered by a variety of alternative perspectives. Public deliberations reveal differences of opinion and compel policymakers and their advisors to give up their claims to privileged knowledge.

Clearly, the three forms of democracy—representative, participatory, and discursive—are not mutually exclusive. The representative democracy of the early twentieth century was not simply displaced in the tumultuous 1960s. Nor was the participatory democracy of the 1960s dissolved by the discursive democracy of the 1980s and 1990s. The 1960s expanded the pathways of participation by institutionalizing citizen involvement in a multitude of government programs. The government's regulatory sphere was extended and a variety of public interest voluntary associations, such as the Welfare Rights Organization and the National Organization for Women, came into existence. By providing more access for citizens, representative democracy was enhanced. In turn, discursive democracy builds on participatory democracy. It gives publicity to marginalized groups and makes issues of identity more salient. More broadly, a discursive democracy encourages recognition of the importance of language and narrative in how we understand the city and our place in it (Flyvbjerg, 1998; Throgmorton, 1996).

For a discursive democracy to thrive, a wide array of public spaces must exist (Beauregard, 1999; Bender, 2001; Goheen, 1998; Young, 1990). Alone, the offices of constituent services, legislative chambers, and political clubs are insufficient. Spaces of a more public nature are needed. In them, people will encounter strangers, people whom they do not know and whose style of dress, language, and behavior may be foreign. Living together under these circumstances means learning how to be tolerant of others. This provides the basis for empathy and for a willingness to engage with those unlike one's self. That engagement might be overt or as subtle as adjustments in posture, visual scrutiny, or polite conversation; people register each other's presence in numerous ways. These are openings for more extended discussion.

The places where people congregate are places where common concerns can be articulated and debated: hearings held by public bodies and city councils, unveilings of proposed private and governmental projects, citizens' advisory board meetings, neighborhood gatherings, interviews with newspaper and television reporters, chance meetings with neighbors, and encounters at public events. In these places people can protest, celebrate, listen and watch, and declare their allegiance to a government or a nation, to one another, or to a social movement (Beauregard, 1995). Democratic deliberations flourish. Democracy takes visual and collective form, in contrast to the isolation and individuality of the voting booth, letters to representatives, or political caucuses (Sennett, 1999). In such places, publics are created and act. As the historian Thomas Bender (2001, p. 73) has argued, a public exists only when people "propose[s] to do something together."

The settings and reasons for gathering are numerous and varied. Consequently, deliberations take multiple forms. They include the instrumental and linear presentations of policy analysts and planners, the strategic calculations of elected officials, the commentary of public intellectuals, and the personal stories of common citizens, among others. Discursive democracy is both substantive and contextual and this embeddedness gives public stories a sense of purpose. It makes them intentional and political (Throgmorton, 1991). Told stories are not simply metaphorical, they reflect on material conditions and the possibilities for action (Garber, 2000).

An urban citizenship is inconceivable without an array of such deliberations and public spaces (Beauregard and Bounds, 2000; Dagger, 1997; Holston, 1999; Isin, 1999). The urban citizen is one who is engaged in an ongoing discourse with those who live, work, and frequent the city. This can only be done where people congregate. The publicness of their deliberations establishes the basis for a vibrant democracy in which politics and policies, governments and their representatives are scrutinized.

The ideal urban citizen, then, is an active storyteller who is expressing a view of the world from a personal perspective. She or he is also an active listener to stories told by others. While public spaces might lend themselves to political speeches, harangues, and avant-garde ravings, the basic democratic work is only done when people interact with each other in ways that allow specific experiences to be set against other specific experiences and to be considered, validated, and challenged. Telling and knowing are connected. Stories are told not just to express understandings and intentions to their listeners but also to reshape them.

A discursive democracy has to enable private stories to become public. To do this, it has to encourage storytellers to consider alternative understandings and to differentiate among personal responsibility, private interests, and public concerns. For this to happen, trust and reciprocity must be strong. Deliberations must be nonthreatening. In the absence of such conditions, citizens will remain silent or defensive.

Of course, not all storytellers will be learners and not all stories will be transparent. Some stories will manipulate and deceive; some stories will be crafted (that is, told strategically) to serve interests that are not fully revealed. Stories that emanate from government agencies, corporate actors, and advocacy organizations are particularly suspect. Having been first formulated to achieve institutional ends, they are less likely to be open to modification than those voiced by less organized groups and individuals.

If deliberations are truly public and knowledge widely available, such stories will be exposed. The storyteller's premises, values, and facts will be probed amidst an interplay of opinion and evidence, thereby revealing the story's intentions. Transparency is the goal. Only in this way can trust be created and people maintain the open-endedness of their comments. If democracy is strong and concentrated power constrained, distortion and manipulation will be minimized (Flyvbjerg, 1998).

The ideal individual story begins as liminal, experiential and expressive, relatively unformed, and inviting. As liminal, stories are sited discursively between personal claims and public significance. To gain legitimacy and support for the story, the storyteller must then transcend personal concerns and connect to a larger number of storytellers. The particular must be made even more meaningful.

Storytellers frequently start by relating their own experiences. This is what they know best and often why they have chosen to speak. In doing so, they unselfconsciously reveal their feelings through their tone of voice and demeanor. The issue is important to them. They are emotionally involved; the issue degrades or enhances their daily life, might influence their future, or violates or supports their values. These are stories about the "city of feeling," not the stories of objective experts relating detached insights about others' problems in a "city of fact" (Rotella, 1998). Expression, though, has its limitations. Sentimentality, moral purity, and raw emotions all weaken the potential for effective collective action.

These stories are often quite spontaneous. They are the stories of people who have not engaged in reflective, calculating revisions of their thoughts before voicing them. What is told is a distillation, sometimes a fragment, of a larger set of experiences and opinions. The stories are in the process of becoming and each telling is somewhat different.

This spontaneity also means that the stories are unfocused and because of this inviting. Storytellers are searching for validation and connection with others who have had similar experiences or understandings. At that point, they are open to criticism and dispute and willing to change their views in the face of stories previously considered irrelevant or unimportant. The intent, as Marris (1997, p. 54) has written, is "to tell stories which preserve some of the complexity and uniqueness of actual events, yet at the same time claim to be typical." Storytellers reach out, trying to connect.

The function of the public sphere is to take such stories and mold them into bases for collective action (Young, 2000). In doing so, individual stories become fewer in number and their commonalities more apparent. Collective action becomes possible. The diversity and tolerance that the city engenders are thus enhanced by a democracy that enables people to engage each other in the city's public spaces. Democracy thrives when that engagement generates varied, dense, and inclusive public understandings. In the public spaces of the city, stories create publics and by creating publics build democracy.

Sustainability

Why, though, must a sustainable city be a democratic city? And how does storytelling fit into this equation?

Sustainability is the antidote to Faustian self-destruction on the one hand and complacency and indifference on the other (Harvey, 1996; Haughton, 1999; Jacobs, 2000). When progress is pursued in the absence of any consideration of its human costs, Faustian self-destruction is the result. The social theorist Marshall Berman claims that it is endemic to modernization. The "drive to create an homogeneous environment, a totally modernized space," Berman (1982, p. 70) has written, leads inexorably to the tragedy of development. In taking control over the world in this way, in attempting to impose our intentions impersonally through large institutions and without regard to history or the particularities of places and cultures, we do damage not only to those in the path of progress but to ourselves.

In the realm of development planning, Faustian self-destruction is associated with a top-down approach that imposes modernization on so-called backward or developing nations. Large-scale electrification projects, industrialization, the mechanization of agriculture in order to expand crops for export, and rural-to-urban migration policies signal an indifference to indigenous cultures, political traditions, and social relations (Scott, 1998). More recently, structural adjustment policies imposed by international aid organizations that require governments to privatize public services and deregulate economic activity have risen to the top of the development policy agenda. These policies have been equally destructive of societies and their environments. Moreover, the modernist fascination with measuring development in economic terms has kept the development community from recognizing the costs associated with such policies (Sen, 1999).

For residents of the United States, sprawl is the current version of perverse development (Rusk, 1999; Wasserman, 2000; Whyte, 1958). Recently revived as an object of public debate and concern, sprawl stands for (in all but the most conservative circles) a waste of natural resources. Sprawl evokes a limitless frontier, a mentality that takes abundant land for granted and assumes that natural resources are infinite.

For those architects and planners who champion multiuse, small-scale, and villagelike developments in the suburbs—the New Urbanists (Duany, 1998)—sprawl destroys community and damages nature. The intertwined quests for privacy, autonomy, open space, and the good life—the relentless American nostalgia for the small town—result, instead, in public problems.

Sprawl has replaced urban renewal as the prime example of the Faustian self-destruction of the city. Under urban renewal, we had to destroy the city in order to save it. The abandonment of the city by industry and commerce led to blight and slums. Demolition and site clearance were then used to set the stage for new construction on land now stripped of its historical heritage and identity. Today, numerous toxic sites produced during the country's rise to world industrial dominance provide a striking example of industrialization's lingering costs.

Sustainability is also intended to save us from complacency and indifference—the disinclination to act in defense of our shared fate. The specific complacency that irks those who embrace sustainability involves threats to the environment and thus to the ecological relationships that enable animal and plant life, and human society, to provide the conditions for a livable future. To be complacent is to ignore the ever-increasing growth rate of the world's population, the encroachment of settlements on deserts and forests, the pollution of the seas, acid rain and snow, the destruction of plant and animal communities, the collapse of fish stocks, and the depletion of spaceship Earth's ozone layer. Instead, those who are self-satisfied place their hopes on the regenerative powers of natural systems or the sensibility of humans not to exceed reasonable limits. Or, they view such destruction as the unavoidable costs of progress. From the perspective of sustainability, this is simply foolish.

To embrace sustainability is to be sensitive to the threatening consequences of action and the disastrous consequences of inaction. Sustainability implores us to act with respect toward nature and to pay close attention to the dangers of inattention. Sustainability addresses our ever-emergent capacity to act together for the common good.

Described in this way, sustainability appears as a defensive practice, a position of resistance to potentially destructive actions. Such a stance can be interpreted as conservative, almost romantic in its wistfulness for a lost natural world. It can also be understood as taking responsibility, a form of stewardship. As stewardship, sustainability encourages people to be responsible for the environment, other living organisms, and future generations.

More than a concern with the natural environment, sustainability is situated at the intersection of environmental protection, economic growth, and social justice (Campbell, 1996). Sustainable development is meant to reconcile these three conflicting interests, a difficult task under most circumstances, but not impossible. In addition, it emphasizes the concerns that people hold in common (such as their shared need for clean water) and the needs of future generations. Sustainability is "organic" in the first sense, seeing the public interest as independent of utilitarian calculation. In the second sense, it urges us to interpret our responsibilities in broad historical, geographical, and ecological terms—responsibilities to future generations, to people and places throughout the world, and to other species (Haughton, 1999; NSF Workshop on Urban Sustainability, 2000).

A sustainable city, then, is one in which environmental quality, economic growth, and social justice coexist and guide public deliberations and actions. In order for these conditions to be realized, the city must be governed in a way that is attentive to the shared concerns of its people and to the future implications of present actions (Harrill, 1999). Initiatives by corporations, governments, nonprofit associations, or individuals that hinder environmental, economic, and social justice or burden future generations must be prohibited. Decisions that improve the welfare of all or of those currently disadvantaged, without violating the welfare of those not yet born, are the essence of social justice (Rawls, 1971). Resource use and economic growth must meet standards of justice and be attentive to the stewardship implicit in the ethos of sustainability.

These goals are both laudable and in conflict with each other (Lake, 2000). Environmental quality and economic growth are often viewed as incompatible, though the relationship is more contingent than necessary. Moreover, sustainability can easily become a politically inclusive way to accommodate growth, discarding environmental limits for a naive belief in the possibility of win-win outcomes (Torgerson, 1999; Wolch, Pincetl, and Pulido, 2002). Growth and justice are in a similar relationship, while the concern for future generations is often displaced by the goal of managing growth.

At the core of sustainability is the need to pay attention, negotiate conflicts, and engage the complexity of the sustainable city. Where democracy is stunted, decisions and actions that weaken sustainability are more likely to occur. When political power is concentrated and unaccountable to the citizenry and when businesses and voluntary associations are allowed to pursue their interests unrestrained by either government or the weight of public disapproval, investments and resource use favor elites over the masses. Damage is done because sanctions against or punishment for such decisions is weak or nonexistent. The costs of economic growth and the pursuit of the good life for a minority of citizens are not part of public decision making. Consequently, environmental quality and social justice are devalued.

Under such conditions, elites acting in their own interests can easily isolate themselves from the damage that they do (Caldeira, 1996). If their corporations pollute rivers, they can build vacation homes on the seashore. If they own most of the country's wealth while others live in poverty, they can avoid the "unfortunate" by traveling in private jets or limousines. If the governments that support their interests use repression to control the masses and crime and disorder are the result, they can retreat to gated communities. If the city becomes unsustainable, they can move to rural areas or live in another, less damaged country. Accountable neither to democratic institutions nor to democratized citizens, elites can do more or less as they please. The result has most often been environmental destruction, economic growth that depletes the resources available to future generations, and inequality.

Democracy provides a number of antidotes (Lindblom, 1977). First, it provides for accountability via a government beholden to the electorate and to public opinion as expressed through a free press. Neither corporate nor government actions can proceed unmindful of the responsibility to act in sustainable ways. Any action that can be cast as public becomes a legitimate subject of political deliberations.

Second, a democratic society provides vigilance. A free press, the requirement that government actions be transparent, and the existence of various planning and policy mechanisms that oversee public actions allow those actions to be scrutinized. Active citizens watch over their communities, the public interest, and the nation's interests. Such scrutiny is essential for accountability. Once they became public, the desirability of the proposed actions can be evaluated and mechanisms of accountability created to prevent or compensate for unacceptable consequences. Accountability, transparency, and vigilance guard against detrimental actions.

Why might citizens concern themselves with public actions or consider the fate of future generations? Democracy can take many forms, and might well appear (as it so often does) as a defensive and begrudging pluralism, with each group mainly protecting and enhancing its own welfare (Weir, 1994).

One answer is a utilitarian calculus that defines the public interest as the outcome of the pursuit of individual interests. That pursuit occurs within a social matrix in which each individual must adapt, more or less, to resource constraints and the needs and desires of others. This mutual adjustment—the political counterpart to the economy's hidden hand—is what produces the public interest (Taylor, 1998). By definition, the resultant combination of outcomes reflects what citizens hold in common. Future generations are only served, though, when utilitarianism is augmented by altruism; that is, when some people espouse interests that include future generations. Altruism is severely constrained, however, when it is reduced to a particularistic interest. Collective responsibility for sustainability is abandoned. If no individual finds sustainability to be important, the logic suggests, society is free to ignore it.

We could solve this dilemma by simply investing a central body with the responsibility to enforce sustainability. This has two shortcomings. First, it invests that body with powers that detract from democracy; centralized bodies minimize accountability in practice and also discourage participation and discursive involvement with citizens. Second, in a weak democracy, any central body is easily captured or constrained by powerful groups and coalitions. The likely outcome is the undermining of support for sustainability.

What we need is a solution somewhere between these two extremes. This is where a discursive democracy becomes important. Neither a representative nor a participatory democracy has the intrinsic potential to generate groups that are committed to sustainability. An elected representative would either have to be altruistic (pulling us back to the individualism of utilitarianism) or be accountable to a group advocating sustainability. The existence of such a group is not ensured in a representative democracy.

Neither is sustainability an inevitable outcome of a participatory democracy, although participation is certainly important in supporting it. A participatory democracy centers on interests; it is people with shared interests who become involved. Those interests might or might not include empathy for the needs and desires of others and thus might fall short of what is required in a sustainable city. The problem is with the interests themselves. The interests are individualized first and subsequently articulated as a group interest, which is always a problematic sequence (Piore, 1995).

A discursive democracy provides a way to achieve sustainability without having to rely on a utilitarian calculation, altruism, or the availability of representatives who just happen to find sustainability politically advantageous. A discursive democracy is one in which people's interests are formed through talk and deliberations (Habermas, 1996a). People's interests are not presocial, but emerge out of social interactions. People reveal their feelings about an issue, listen as others speak, reflect on what they have said and heard, and search for common ground. Out of this interaction arise interests that are intrinsically collective, not predetermined or the aggregation of individual interests. Storytelling is very much a part of this discourse. It is "often an important bridge . . . between the mute experience of being wronged and political agreements about justice" (Young, 2000, p. 72).

How, though, does sustainability become important in such a democracy? Simply, it emerges from the recognition that many interests depend for their satisfaction on shared conditions. It requires, as James Throgmorton notes, acts of environmental imagination. In the absence of these conditions, few interests can be realized for most people. The existence of a nondiscriminatory labor market or efficient and ecologically sound transportation are but two obvious examples. Once grasped, this point of view encourages people to reflect on their commonalities, their intersecting histories, and the way in which society is organized, or not organized to address their concerns. People begin to see that preserving the environment, managing economic growth, and achieving social justice make for a good society, one that is better equipped to meet their needs, desires, and interests.

Such an understanding also encourages them to consider their collective future. The present is not the totality of their concerns. Most public deliberations have to do with how and when a burden will be lifted and lives will be made better, or how society can cope more effectively and justly with discrimination, injustice, and inequality. It is only a short step from making the future part of deliberations to thinking about future generations.

All of this can emerge from and be supported by storytelling. The process of creating sustainability requires public deliberation; sustainability requires democracy. A healthy democracy, in turn, requires widespread engagement and public deliberations. And public deliberations require storytelling. When people can share their stories and negotiate their understandings and interests, they are more likely to be concerned with the sustainability of their shared world, now and in the future.

Conclusion and Caveats

In a perfect world, a discursive democracy would produce sustainable cities. It would anchor participation and representation and turn attention relentlessly to the need to protect the natural environment, guide economic growth, and pursue social justice. Citizens in a wide variety of settings and from the full array of social positions would tell stories to each other. The result would be a foundational discourse in which issues of sustainability would be unavoidable. United in a shared fate, citizens would demand that their institutions address collective and future needs.

Storytelling is central to all of this. It is essential to a robust democracy and is one of the most hopeful paths to the sustainable city. Storytelling alone, though, is insufficient. The state must be kept democratic and citizens must remain vigilant.

My argument, of course, is highly normative and suffers from all the optimism that hope carries. Because it recognizes the barriers to sustainability and democracy and the problematic link between storytelling and sustainability, though, it falls short of utopianism. In an imperfect world, the forces that threaten sustainability are powerful, and numerous counterstories are mounted to undermine the ability to imagine a sustainable city. That only experts know how to manage the environment and that sustainability will reduce prosperity are two of many examples. Stories are meant to discipline our understandings. Thus it is inevitable that public stories will be used to control others. Yet, even those who enter the public sphere strongly committed to their positions and interests are susceptible to counterarguments. Public ideas have a force, a power, that is itself not easy to control (Kelman, 1988).

While the city might well be the site of public deliberations about its sustainability, sustainability is not only a local process. Sustainability is multiscaled; its consequences pay no heed to political boundaries and its attainment requires action in local, regional, national, and global arenas. Consequently, the discursive democracy that supports a sustainable city must itself be multiscaled. It must be capable of knitting together stories from various locales and mounting actions simultaneously in a variety of arenas.

All of this, and more, constitutes the challenge of the sustainable city. To move closer to that goal, we must nurture a democracy of robust and inclusive storytelling.

II
Raising Questions/Raising Cain

The three essays presented in part I offer a set of ideas about how story, sustainability, and democracy mutually construct one another and American cities. They also provide terms and tools that readers can use when reading and interpreting the essays and stories that appear in part II. Those terms and tools suggest a set of questions that the reader can, in the interest of sustainability, ask of the texts that follow or indeed of any story, essay, or plan that attempts to address the needs of the American city. Initially, we offer the questions clustered in the categories of story, sustainability, and democracy, but in the end these categories best serve urban theory and practice if the reader understands them as integrated.

What's in a Story?

Some of the essays in part II use conventional linear narrative to structure their content; some use multiple or fragmentary or spatialized narrative pieces to present their purposes; and others are largely exposition about narrative as it works with other devices to explain the work of urban practitioners and the needs of America's cities. These textual differences matter. However, whether narrative acts principally as the form or the content, the medium or the message, each essay's use of narrative may be usefully interrogated with questions derived from part I.

Begin, then, with the recognition that each essay is an artifice, a thing the author has constructed for a purpose. Who *is* this author? What gives him or her the authority, the legitimacy, the status to speak? What about the "speaking" voice of each text? From what professional and experiential frame of reference does the voice speak? In fact, quite literally, from what temporal and geographic location is each voice speaking?

What is the author's purpose? How has the author chosen to structure the essay in order to promote that purpose and convey that meaning? Consider, for example, these questions: How does the text begin and end? In narratives per se, what moments in real-life events has the author selected to serve as the beginning and ending of the text? In addition, which life events does the author select for inclusion in the narrative or essay as a whole? Among these events, which are given the greatest space (the longest duration) in the text, either all at once or in frequent repetition that falls into a significant pattern? On what territorial space at what geographic scale do the author and text focus? What dominant images, chronotopes, represent the intersection of time and space in the text: the road, the threshold, the neighborhood, the park, the meeting room, the court room?

What kinds of "characters" populate the text? Are they abstract objects (i.e., concepts) common to exposition and argument, such as narratives, models, theories, information systems? Are they representations of human types, such as technical experts, scholars, residents, members of a particular racial or ethnic grouping or socioeconomic class, citizens, eligible voters, elected officials, property owners, consumers and producers, stakeholders? Or are they depictions of individuals, such as authors themselves, named office holders, named activists, pseudonymously named defendants, named scholars? What differences do these choices make? Which of the characters populating this place have sufficient territorial and textual space to articulate their points of view? Does the text allow multiple voices to speak virtually at once or does it control the conversation so that one or two voices dominate? Do the speaking voices all mark territory as their "turf" or do some offer other possible relationships between human communities and space? Do the chronotopes of different voices interact dialogically in the text so that their real world dialogue is imaginable? Which characters occupying the territory remain silent in the text?

Having considered how the essay presents its author and constructs such textual features as duration, frequency, voice, characters, and chronotopes, observe how the essay constructs, or conscripts, the reader it imagines for itself. Does it directly or implicitly address "you"? What values and desires does it ascribe to this "you"? Are they your values and desires, you the actual geohistorical reader, so that you fit comfortably (is that fit too comfortable?) into the stories and purposes of the essay? Out of your own authority, legitimacy, status, habits, or taste do you make a preemptory decision that some essays and stories have something to teach you and others do not? Do some essays compel you to suspend your habits of thought and go along for the illuminating surprise? How do they do that? Do others conscript a reader so antithetical to your values and desires that you resist the purpose of the text and interpret it against the grain? Can your resistance be enlisted, with that of others, to renew rather than suppress the purposes of the text?

Narrating the Sustainable City

All of the questions about the forms of the essays generally and the narrative particularly may be read with one or both eyes on sustainability. To these questions we add others that are specifically focused on the goal of sustainability.

Consider, for example, an essay constructed as a linear narrative unfolding over time. Does it intersect this diachronic sequence of events with the synchronous occurrence of events in other locations, which disrupts the very linearity that seems to hold the narrative together? What do such spatial disruptions of temporal form tell us about the tenticular radiations and remote effects of the issues at hand, and hence about the boundaries of "our place"? How does each of these essays handle the relative powers of temporal order and spatial order? Does the story imagine cities and other human settlements to be part of a larger regional or global life space shared with other living creatures, or does it treat them as independent sites of human agency, rather like air-conditioned office towers in which "important things happen"? How might a different spatial frame or geographic scale alter the meaning and effect of an essay? Does the essay articulate an interaction of different scales?

Ask yourself how the essay constructs who "we" are. Are other living creatures, complexes of creatures (such as watersheds or forest ecosystems), or naturally occurring events and processes (such as floods, tornadoes, hurricanes, and earthquakes) foregrounded as prominent characters, or are they relegated to the deep background, where they provide nothing more than a stage or setting for human action? If the essay puts these nonhuman creatures and processes in the foreground, does it treat them as somehow pristine, untouched by human hands, or does it articulate interconnections between them and the human occupants of places? Does the essay treat these nonhuman beings, complexes, and processes as part of the "we" who reside in "our place"? Does it make space for diverse stories about what it means to reside in that place?

Finally, consider the ways in which the essay incorporates an environmental justice perspective into its articulation of community. In what ways are unpriced costs (e.g., global climate change) and remote effects (e.g., exported abuse of labor) made visible in the essay? Do wastes (often hazardous) figure prominently as features of the narrative's setting, and thereby signal that the place is at the receiving end of hazardous and wasteful economies, or does the narrative implicitly export those wastes to other places and other narratives?

Enacting a Discursive Democracy

Just as multiple stories are a constitutive, albeit far from simple, part of imagining sustainable cities, democratic engagement is equally necessary to build and maintain them. The essays printed here, in fact all discourse, should be answerable to the complex demands of sustainability and its foundations in democratic practice. To what extent and in what ways does each of these essays encourage you to recognize and engage strangers (other actual geohistorical readers), and hence facilitate the forging of trustworthy agreements between the author and readers and among readers? Alternatively, to what extent and in what ways does the text exclude, repress, or marginalize counternarratives, in the name of accepting the outcomes of majoritarian electoral politics and representative democracy, enhancing market competitiveness, deferring to technical expertise, maintaining the investment of local elites, or for other reasons?

How does the reading of one text here influence your interpretation of the others? Does the juxtaposition of these essays and their stories alter their individual meanings? Does it affect your understanding of how story, sustainability, and democracy mutually constitute one another and together can transform American cities now so largely privatized and segregated?

Finally, in what ways do the texts presented in part II exemplify, amplify, modify, or challenge the key themes and arguments presented in part I? We invite our readers to hold our feet to the fire.

Editors' Introduction to Chapter 4

With an ear for good story and an eye for the messy detail of urban realities, Rotella here extends his earlier argument (1998) about the reciprocal relationship between the "city of feeling" as expressed in such venues as literature and the "city of fact" as built in neighborhoods and negotiated in political forums. The old neighborhood in question is the South Shore of Chicago's South Side, the subject of much discussion and the object of much anxiety and reforming zeal. If apocalyptic stories predominate among those told about the South Side, Rotella treats them not only as dire predictions about the future but also as cautionary tales advising neighbors to behave well in a precarious present. In Rotella's hands, neither neighborhood nor story are rigid or sentimental objects, as Eckstein fears they often are.

In addition to the literature Rotella addresses, St. Clair Drake and Horace R. Cayton's Black Metropolis *(1945), Arnold Hirsch's* Making the Second Ghetto *(1983), Alan Ehrenhalt's* Lost City *(1995), and LeAlan Jones and Lloyd Newman's* Our America *(1997), provide evidence that the South Side, generation after generation, is a signal neighborhood—a beleagured place in pursuit of a sustainable vision. Reports of its death are premature.*

"What's the story here?" If we can begin to answer that satisfactorily, . . . [we] can make decisions out of compassion rather than fear.

—Joe Barthel

4

The Old Neighborhood

Carlo Rotella

Neighborhood, as a concept, is "hard" in the sense that a neighborhood is a physical artifact, a bounded and built-up space containing people, money, buildings, and other elements that can be measured with numbers.[1] I grew up in South Shore, a largely residential neighborhood on Chicago's South Side about 9 miles south of downtown. Jackson Park and 67th Street form South Shore's northern boundary, Stony Island Avenue the western, 83rd Street the southern (in most authoritative accounts), and Lake Michigan the eastern boundaries. About 57,000 people live there, divided into about 23,000 households. Recent tabulations indicate that almost 5,000 of these households have an annual income over $50,000, more than 10,000 have an annual income between $15,000 and $50,000, and more than 7,000 have an annual income under $15,000. At least 97 percent of the neighborhood's residents are officially identified as black, a soft cultural distinction that commands the status of hard fact in law, governance, business, politics, and social arrangements. There are shopping strips along 71st Street, Jeffery Boulevard, and a few other major streets, but houses and walkup apartment buildings predominate. More than half of South Shore's generally well-worn but solid building stock predates 1950.[2]

Neighborhood is "soft" in the sense that it describes a quality of civic life and of inner life, a feeling of relation to people and place, that is sustained or destroyed through the statistically unmeasurable processes of culture. To think of someone—even an enemy—as a neighbor acknowledges an obligation or regard not always extended to strangers; to think of a piece of the city as one's neighborhood is to acknowledge an investment in it that goes beyond rents or mortgages.

The terms and potential consequences of that less tangible investment can be found in the ways people act toward one another, the ways they imaginatively inhabit the landscape, the ways they think and talk and write about their neighborhoods, the stories they tell. Much of the foregoing is irrecoverable; some, especially the small fraction put in writing, can be partially recovered.

South Shore's literature—which, as you will see, I define loosely enough to include novels, essays, journalism, sociology, criticism, advertising, letters, comedy routines, and perhaps dreams—provides an occasion to consider the relationship between the soft quality of neighborhood as an aggregation of feeling and the hard facts of neighborhood as a composite artifact. My subject is the sustainability of neighborhood, and I put my questions to the writing of South Shore. What does the literature suggest about how a neighborhood (hard) generates and sustains the quality of neighborhood (soft)? What does the literature suggest about how that quality of neighborhood (soft) in turn helps to sustain or destroy the social, economic, and physical arrangements that add up to a neighborhood (hard)?

For the past century, South Shore has been a consistently respectable district that attracts enthusiastic, ambitious, upwardly mobile first-time homeowners. So what does it mean that its literature has insisted on imagining the *un*sustainability of familiar orders and ways of life? One of the literature's most striking features has been persistently recurring visions of imminent neighborhood apocalypse.

This essay, then, considers a local instance of the larger relationship between what I have called elsewhere the city of feeling (constructed in words and images) and the city of fact (made of steel and stone, inhabited by flesh-and-blood people).[3] I am also talking about my old neighborhood, though, so I will observe the three unities of traditional Chicago storytelling: there will be railroads; there will be winter; there will be sentiment and market logic in close, mutually conditioning conjunction. And, since the quality of neighborhood can persist in one's inner life long after the bounded space of the old neighborhood has been left behind (just as, to reverse the mismatch of feeling and fact, one can persist in living next door long after having mentally disinvested from one's neighbors), there will be bad dreams.

71st and Jeffery

WFMT, the insufferable Thinking Person's Radio Station to which my parents listened when I was a kid in the 1960s and 1970s, used to play a recording called "The Chicago Language, Tape 2, Side 1: Street Names," in which a comedian gets laughs by pronouncing street names. It is the essence of local humor, a map made of inside jokes that are funny to the extent they give continuing life to neighborhood stereotypes while simultaneously mocking them. A derelict souse says "West Mad—*hic*—ison"; a dignified southern basso, enunciating every syllable of a hard-won name change, says "Doc-tor Martin Luther King Jun-ior Mem-o-rial Drive"; and "Meigs Field," drawn out and dopplered, becomes the

Figure 4.1
The northwest corner of 71st and Jeffery in 1955. At far left is the entrance to the Bryn Mawr train station. Behind the station and tracks is the South Shore Bank. (Source: Reprinted with permission of Chicago Historical Society, negative number ICHi-34360; photograph by J. Sherwin Murphy.)

sound of a small plane missing the runway and hitting the lake with a glass-of-water-sized splash. The comedian does a South Shore joke, too: a terrified white voice, out of breath from running or hysteria, gasps "71st and Jeffery!" Many of the other jokes have lost their charge or their referent over the years, but "71st and Jeffery!" still makes people I know laugh, not because it describes a continuing social reality of pandemic violent crime against white victims, but because it both articulates and lampoons a perception of South Shore as white people's worst nightmare. That perception achieved citywide circulation as received wisdom and acquired seemingly eternal life when almost all of South Shore's white residents were getting out in the 1960s and early 1970s. Even then, when street crime did rise dramatically, and even now, when South Shore has been a black neighborhood for most of two generations, if you were to hang out at 71st Street and Jeffery Boulevard on any given day you would see white people going about their ordinary business among the black majority, unremarked upon and unmolested. But the joke, which is not so much a joke as simply a way of saying two street names, hangs primarily on what people *say* about South Shore, the combined social and cultural history of which is both brutally compressed and astonishingly well expressed by those three words.

The old Illinois Central (I.C.) tracks run down the middle of 71st Street, South Shore's main east-west corridor. The tracks are at grade, the way railways passed through neighborhoods in the late nineteenth century and early twentieth century, when Chicago grew so quickly around its rail lines. Because the tracks are not elevated the way they so emblematically are in other parts of the city, all traffic across 71st comes to a halt when warning bells and lowered crossing gates signal a train's approach. My family moved to South Shore in 1967, when I was 2 years old, and I lived there until I graduated from college in 1986; it seems to me, in retrospect, that I spent a significant part of the 1970s waiting for the I.C.'s warning bells to stop or start ringing. Either I was stuck at a crossing while a train passed or I was skulking on the platform of a station, hoping that a train would get to me before the cold or the pea-coated, sideways-Pittsburgh Pirates cap-wearing thugs of the era did. *You* know: "71st and Jeffery!"

South Shore originally grew from the bud of the Bryn Mawr station at 71st and Jeffery in the late nineteenth century; the railroad and then the

Columbian Exposition of 1893 drove development. Today, that station's raised wooden platform commands a view of South Shore's busiest intersection. Cars pass on all sides, kicking up slush in the wintertime. Pedestrians go in and out of stores and line up at bus stops. Kids patrol in search of action. The last time I went back to visit, customers were emerging from Italian Fiesta, as they had in my childhood, bearing pizzas of preternatural thinness—designed, my brother Sal theorized long ago, to fit through the narrowest possible slot in the bulletproof glass. From the Bryn Mawr platform you can also see—in your literary mind's eye—Studs Lonigan and Gayle Pemberton.

Studs Lonigan, the protagonist of James T. Farrell's trilogy of novels published in the 1930s, lived on Jeffery just south of the intersection with 71st, if I read the novels' imagined landscape right. Studs is a prewar incarnation of a universal type, the sucker-punching neighborhood thug who lives for the approval of the pack. Studs runs with the Irish crews that used to make black people say "71st and Jeffery!" themselves once upon a time. However, give Studs a makeover and an updated vocabulary, including gestural innovations like the repertoire of finger-throwing gang signs that became so popular in the 1970s, and he could pass as a semicommitted member of the Gangster Disciples, Vice Lords, or other latter-day black gangs of the South Side.

Uplifting though it may be to realize that all punks are brothers under the skin, hands figuratively joined in a cross-generational and cross-racial tradition that links antediluvian rushers of the growler to contemporary streetcorner pharmacists, Studs also has a particularistic role to play. He is a scrappy but doomed exemplar of the South Side Irish, for whom South Shore was a staging ground on the generations-long journey out of working-class slums and into the suburbanized middle class (as it has been for African Americans after them, although of course the immigrant model does not transfer without significant qualification). Farrell, in keeping with the literary tradition of Chicago realism, bends that story's upward arc into a downward slide toward failure. Studs, embodying his urban tribe, collapses under the combined pressure of large economic forces (especially the depression) and internal contradictions that divide him against himself, especially the tension between aspiring to individual success and refusing to recognize the communal ties and possibilities that inevitably attach him to other members of the working class. Broke, broken, and mortally sick at the end of *Judgment Day*, the last of the trilogy, he staggers off the train at Bryn Mawr and drags himself down the block to die. Like so many characters in Chicago neighborhood novels, Studs, and the way of life for which he stands, has been ground up by the relentless action of business as usual in the city.[4]

Around 1980, a half-century after Lonigan's death march, Gayle Pemberton, who is now a professor of African American Studies and English at Wesleyan University (the college on a hill in Middletown, Connecticut, to which I went when I left South Shore), also finds her way to 71st and Jeffery in her essay, "Waiting for Godot on Jeffery Boulevard." She describes a Saturday morning walk on 71st Street past heaped garbage, shuttered storefronts, a few surviving businesses, and menacing male idlers. This once-booming shopping strip never recovered from the sudden withdrawal of money, opportunity, and power from South Shore in the 1960s. Reaching Jeffery, where there are more signs of life, but also more young men "with plastic bags on their heads" messing with passersby, Pemberton enters a shoe repair shop that might be on Studs Lonigan's old block. Inside, a group of middle-aged men are extolling the verities of the Book of Genesis. She wonders how they can talk so complacently about God's creation when "just outside the door was desolation and death." She wants them to "act in the face of the ironies of black American life, to leave Genesis and the *fait accompli* of the Earth's formation behind, to stop preaching to the converted and get out in the streets to do some small thing, like suggesting to young men that obsession with one's genitals stunts one's growth and that curls, though no doubt pleasing to their wearers, look like conked, greasy Afros to a whole lot of people—potential employers, for instance." Pemberton makes 71st and Jeffery into ground zero of a biblical devastation visited upon the black inner city, the kind of plague that drives young men violently mad and turns old men into maundering weaklings.

The essay proceeds to other matters, especially the deaths of men—her father, Paul Robeson, a friend who died of AIDS, a friend who died of alcoholic despair—and a woman's near death: her mother has been stabbed by someone who broke into her house in Kansas City. Enough is enough; the mother is moving. At essay's end, having returned through the annihilating landscape to her sister's apartment on South Shore's lakefront, Pemberton concludes that "something out there today is too much for me," puts her head back, and cries herself to sleep.[5]

I do not recognize my old neighborhood in Farrell's or Pemberton's death-seeking visions of decline and fall, but I do recognize the apocalyptic tone shared by the stories. Endowed with a healthy complement of strivers and good neighbors, South Shore was not and is not a bad place to live, but the neighborhood's air of precarious stability in the face of larger forces does seem to inspire a doom-haunted temperament. Stories about South Shore are always alert for a whiff of judgment day hanging over blocks of bungalows kept trim by owners who struggle to make good against the grain of opportunity. If you look for it, you can find a Gothic quality in the big houses of the Highlands, South Shore's fancy district, fortified like a string of frontier outposts. Entropic anxiety radiates from the depressed main drags and public housing where respectable poverty coexists with the more hopeless kind. As a real estate man observed to an inquiring reporter in 1969, "The world is always coming to an end in South Shore."[6]

He did not mean that South Shore is in a constant and extreme state of social trauma, because that is not the case. The neighborhood has been blessed not only with enthusiastic homeowners but also with the central presence, right at the corner of 71st and Jeffery, of the internationally celebrated South Shore Bank. Since being taken over in 1973 by a group of socially committed directors, this powerful institutional proponent of "community capitalism" has dedicated itself to helping provide the material basis of a viable inner-city urbanism. The bank and its affiliates, for instance, helped rehabilitate 12,000 apartments in South Shore in the 1980s and 1990s.[7]

The bank did not single-handedly save the neighborhood, as it is often credited with doing. In fact there are some who claim, conversely, that during the 1970s the bank's influence hastened the collapse of remaining community networks by converting the project of saving the neighborhood from a civic to a fiscal footing. But the bank's commitment to South Shore did make it easier for individuals, community groups, businesses, and government to invest in the neighborhood. Those investments built links of capital, joining past to future by way of a viable present. Gayle Pemberton walked right by the South Shore Bank on the way to her encounter with what she calls "desolation."

When I asked Pemberton during a 1998 interview about the bank's invisibility in the landscape of "Waiting for Godot on Jeffery Boulevard," she said it had not been the result of an oversight. She knew the bank well, but she remained unimpressed by its commitment to the neighborhood. "My sister worked there," she said, "and she was not treated well. The bank is a cold place, because money is cold. The bank maintains itself, whatever happens to the neighborhood." In that sense, the South Shore Bank might have made the perfect centerpiece for her essay's cold, sinking South Shore. "I felt that South Shore is a neighborhood by territory, but it has no feel of a neighborhood. Where would you put the town green in South Shore? It's split like a zipper by the tracks. I've lived in places where a street like 71st will have a set of players you recognize, but it just never felt that way to me. I never saw people. It was a place where life was lived indoors, where the curtains were heavily drawn. There was a fearfulness out of proportion with what felt scary in the street. The blinds were drawn to their own safe neighborhood." That last image captured the essence of South Shore for her. "It's about holding on tenaciously to a class place," as opposed, in her view, to a collective sense of community.

The real estate man who observed so astutely that the world is always coming to an end in South Shore meant it as something like literary criticism, an interpretation of the stories people tell about their neighborhood: stories about individual ambition and desperation caught up in grand-scale movements of people and capital, about the attenuation of neighborhood bonds by centrifugal forces as varied as racial antipathy and economic transformation. And he meant, especially, all the stories of succession and decline told by residents in response to newcomers. "You know," he told the reporter in 1969, "when the Irish came to South Shore, the English thought the end of the world had come. When the Jews came, the Irish thought the end of the world was here. Then when the blacks started coming, there was the end of the world again."[8]

I would add to his list the more recent reactions of some established black residents to new, poorer black residents who have been displaced by the dismantling of high-rise housing projects in other parts of the city and relocated to South Shore's scattered-site public housing. "It is as if the gates of Hell . . . opened, and these people were let out," wrote South Shore resident Hattie Wilburn in a letter about the newcomers in 1998.

"I had to ask again, where did these people come from? And, lo, I was told they came from the projects, the CHA [Chicago Housing Authority]. And as they tear down more of these projects, we can expect more of these people to be relocated in our neighborhoods."[9]

She was wrong about those particular troublemakers (they had not come from the projects), but she was retelling the root South Shore story with unerring precision. Once again, at century's end, when large parts of the neighborhood were in much better shape than they were during the rocky times of transition from the mid-1960s through the mid-1980s, people in South Shore were getting up at community meetings to tell horror stories of a near future in which barbarous invaders would obliterate South Shore's property values, schools, and networks of community in a wave of ignorance and criminality. In other words, the world would be coming to an end once more in South Shore. And once again the threat of apocalypse proceeded from the fit between the precariousness of individual class status and the scarcity of resources, both material and psychic. Once people decide that they cannot afford to invest in a neighborhood, that world begins to end.

The Old Neighborhood

South Shore, which is just a place where people live, is not famous or important in the grand scheme of things. But plenty of people know plenty of places like it, and South Shore stories speak to them, not only because these stories refer to a familiar social reality, but because these stories enact a familiar urge to turn the dull facts of daily life into operatic narratives brimming with drama and meaning. Perhaps that explains why South Shore has enjoyed national circulation in recent years via two meditations on the meaning and function of such neighborhood stories, both entitled "The Old Neighborhood." People in certain parts of New York City or Los Angeles may have grown used to seeing their neighborhoods on the screen, on stage, or in print, but people from places like South Shore still overreact to it because for at least a moment the world seems to acknowledge that one's old neighborhood and its lessons really might matter to everybody, not just to the relative handful of people passionate about a particular collection of streets, buildings, and neighbors.

South Shore turns up in Ray Suarez's *The Old Neighborhood: What We Lost in the Great Suburban Migration, 1966–1999*, a memorial to values associated with the industrial urbanism that took shape around factory, ethnicity, and church and was dismantled in the past half-century by deindustrialization, suburbanization, urban renewal, and racial conflict. The capitalized term "Old Neighborhood" stands for this declining way of life and for corollary certitudes and virtues one might perceive as similarly swept away by forces labeled "postindustrial," "postmodern," and just plain "post-When Things Were Good." Suarez, who used to host *Talk of the Nation* on National Public Radio, claims, "We lost plenty in those years after World War II. People talk about the closeness, the intimacy of the old urban neighborhood. People talk about their friendships found and lost, the adventures of city life, waiting for the old man to come home from work. We knew each other then. We saw our own faces plainly, in the mirror and in each other's eyes." That this sounds like the most fuzzy-minded nostalgia for a fantasized urban village constitutes part of his point: this is how people talk. Suarez warns against "the sleight of mind" by which we turn the urban past into a lost golden age—not only because this trick trivializes the complexity of the old neighborhood, but also because we thereby place the good urban life in a no-longer retrievable past in order to make what he calls "suburban exile" seem " 'necessary,' and unavoidable."[10]

Suarez touches upon the case of South Shore early in his book to demonstrate that sometimes "the forces at work are so big, the geographic and social pressures so strong, that even if you do everything right, it still won't work."[11] Trying to "do everything right" means that in South Shore community groups tried to manage the process of integration so that whites and extant businesses would not bolt en masse, to be replaced by middle-class blacks and then progressively poorer blacks caught up in the familiar cycle of resegregation driven by economic and governmental abandonment. "It still won't work" means that in the end whites and capital did bolt and the neighborhood had to pass through the cycle, although the South Shore Commission and others acted to mitigate its effects by helping to retain the South Shore Bank and some residents who might otherwise have bailed out. Arranged end to end, the titles of sociological studies of South Shore since the urban crisis schematize what one of those books calls "the 'classic' model" of traumatic neighborhood change: *Managed Integration: Dilemmas of Doing Good in the City; Paths of Neighborhood Change: Race and Crime in Urban America; Community Capitalism: The South Shore Bank's Strategy for Neighborhood Revitalization.*[12] The classic model calls for a golden age followed by a decline. Even as Suarez rehearses the familiar tragedies of these first two acts, he reminds us that act three has yet to take form.

If Suarez wants to dry up the often hysterical discussions about places like South Shore and demystify legends of the fall, David Mamet's often hilarious play "The Old Neighborhood" explores our humid, self-dramatizing capacity for processing the facts of urban life into a legend we declaim from our privileged position atop that headstone cum soapbox called "the Old Neighborhood." In the play, a fortyish man named Bobby Gould returns to Chicago and visits with his old friend Joey, his sister Jolly, and a one-time girlfriend named Deeny. They talk about Mamet's usual mess of seemingly tangential subjects—their marriages and failures, who was or was not a fag or a dyke, why their parents hated them, how fine it would feel to be an expert gardener or an old-time Hollywood mogul, what their lives would have been like had their parents never left the Old Country, the topography and meaning of the Old Neighborhood.

The neighborhood in question is Jewish South Shore, which Mamet (born in 1947) knew in its late heyday. As one sociologist describes South Shore in 1960, when Mamet was thirteen, it "was no longer a prime area; it was the old neighborhood for larger and larger numbers of Chicagoans who moved on to the suburbs to raise families or to the lakefront highrises [farther north]. . . . [13] Looking back at Jewish South Shore over a widening gap of social distance and decades, Mamet's characters in "The Old Neighborhood" see desolation: "Oh, Bobby, it's all gone," says Joey, "It's all gone there. You knew that." This comes after an extended discussion of the broadly symbolic Mr. White's broadly symbolic shoe store on Jeffery Boulevard. (What *is* it with shoe-related businesses on Jeffery and the decline and fall of South Shore?) Bobby did know that, just as everybody in the play knows certain routines of nostalgia and recrimination like "it's all gone." The one that goes "71st and Jeffery!" would be another.[14]

As in everything Mamet writes, characterization eclipses circumstances. "The Old Neighborhood" is not really about South Shore, or about neighborhood at all; it is about the dramatization of oneself using whatever props come to hand, family and local history among them. All the various discussions, not tangents but spokes in a wheel, return to the fantasy of living in a way that is somehow more satisfyingly connected to the world and to other people. Deeny imagines that being a gardener would be a way "to *use* the world, I think—those things of the world we could take in: food, or air, you know, and *use* them." She has a vision of herself with cigarette and coffee, looking out a window at her garden. "In this garden there are plants that I have planted and perhaps I have raised them from seeds or cuttings, do you know? The way they do . . . ? . . . They call it 'forcing.' Or they call something else forcing and call this something else."

Joey fantasizes about living in an Old World in which either the Holocaust never happened or has not happened yet, a golden-age shtetl where young beauties bring him fresh-cooked treats and he is respected by fellow men. "They'd say, 'There goes Reb Lewis, he's the strongest man in Lodz.' I'd nod. 'He once picked up an ox.' (*Pause*) Or some fucking thing." All the ellipses and quotation marks, all the stiltedly elevated diction that occasionally breaks down into "some fucking thing" or another when the speaker struggles for the right effect, all that standard Mamet business shows how deracinated characters try to force the ill-fitted, desultory materials of life into blown-up dramas worth enacting: How I Got to Be Me, What it Means to Be Me, What Happened to My Old Neighborhood.

Mamet's characters, in other words, elaborate on Pemberton's feeling that even—or perhaps especially—an apparently passionate commitment to South Shore masks a deeply felt right to disconnect from any sense of collective peoplehood except as a safely distanced nostalgic curio. (The same might be said of similar narratives about other Old Neighborhoods in urban America. We seem to want to blame our shrinking sense of public life on the market or on malign interlopers, and it is an indisputable fact that powerful interests impose themselves on modest communities that cannot mount an adequate defense. However, let us not forget how many neighbors already have their bags half-packed when history, succession, or big capital comes calling.) Deeny comes closest to laying out the stakes of all this bad-faith ritual edifice-building in a typically flailing speech about human disconnection and "the stupid *molecules*." She wonders what it means that experts keep breaking down "Whatever the smallest unit is" into even smaller units, which allows her to bring the play's third and last act to a climax. Bobby, trying to finish a thought about "faith" that Deeny has interrupted, finally gets there by saying that "things will not turn out right" and she responds, "Well, they *won't, will* they? . . . (*Pause*) In the world. The, the, the, world . . . and I was talking about 'faith.' And you say, 'this is ending.' Well, then there's *another* thing. And that will take its place. And sometimes that's okay. But then, sometimes, that's just cold comfort. *Isn't* it?" The tradition of South Shore apocalypticism could not ask for a better, more expressively inarticulate peroration and restatement of a first principle of neighborhood life in the American inner city: if one dispensation is coming to an end, then there will be another to replace it; the world we know will come to an end, and it will not.

Finally, all that scenery-chewing noise about disconnection turns out to be itself the stuff of connection, identifying the characters to one another as members of a lost, diasporic inner-city tribe trying to reconstitute itself. To say "71st and Jeffery!" is not only to rehearse a routinely distancing account of succession and decline, it is also to speak the name of the intersection and the neighborhood with hot, intimate feeling that can bind one to others and not just to the facile memory of a place left in one's socioeconomic wake.

Investments

The potentially binding surge of feeling inspired by neighborhood— Old or otherwise—can be potent in all sorts of ways. It is not just useful for creating good theater, justifying past actions, or articulating a sense of community; the investment or disinvestment of feeling can also be mobilized to facilitate economic behavior, such as selling and buying property.

One of the most heartfelt texts in the South Shore literature, Pat Somers Cronin's newspaper essay "The Agony of South Shore," amounts to a combined call to action and *ego te absolvo* for ethically conscious, racially tolerant white people who have resisted coming to the conclusion that it is time to sell their houses and move out of the neighborhood. Published in 1968, at the height of tension over neighborhood turnover,

the essay reviews the increase in violent crime and racially inflected conflicts that accompanied the arrival of blacks in the neighborhood.
Cronin, writing in the second person, asks like-minded people to recognize themselves as a people, a recognition that will lead them to act
accordingly by moving out while they can. Once upon a time, "you never
thought of leaving"; after all, "You had gone to college with Negroes,
and traveled in Europe with them; they weren't Negroes, they were
friends"; you felt that "of course they were welcome." But purse snatch-
ings, armed robberies, the departure of familiar neighbors and businesses, and a train of large and small indignities have led you to new
conclusions. "Now you don't know," she suggests in the middle of the
essay, but by the essay's closing paragraph, you *do* know: "You no longer
feel you are useful to the interracial effort in Chicago. Your early good
will has been replaced by discouragement; mentally, right now, you are
ready to leave. You are beginning to feel foolish for thinking you could
change a situation. You even begin to acknowledge that those who
moved were smarter. . . . and certainly their children are safer." *Because*
you love South Shore and your children, in Cronin's formulation, you
must recognize that it is time to go. "Is South Shore a better place to live
because you and your neighbors stayed?"[15] The essay expects you to
have the moral courage to answer in the negative, thereby using the force
of your passion for South Shore, judolike, to free yourself from the grip
of the quixotic self-imposed duty of staying there.

Of course, the potentialities of a feeling for South Shore can also be
put to the opposite use: to promote investment and development. One
piece of South Shore literature designed to do exactly that is a brochure
circulated in the late 1990s to advertise "an enclave of elegant new town-
homes" developed on the lakefront at 71st Street by the ShoreBank
Development Corporation, an offshoot of the South Shore Bank.

The development, christened LakeShore Pointe, has one of those
names that is easy to laugh at. The removal of the space between Lake
and Shore (like the missing space in ShoreBank) wants to imply stream-
lined up-to-dateness, as if the development was a particularly sexy website rather than just plain-jane brick and mortar. If they could have
gotten away with calling it eLakeShorePointe.com, they would have. The
extra e does show up, but it is appended to the *end*, to Pointe, a move
conventionally understood to impart an instant Old World classiness

that balances the up-to-date streamlining of LakeShore. The clanking semiotics of developmentese can be painful to consider, but I take the brochure seriously as a piece of South Shore writing that constructs a fragmentary city of feeling from elements of the city of fact—a city of feeling that will, to the extent that buyers take notice, have a significant reflexive effect on the city of fact.

"History," says the brochure, "tells us that South Shore has always occupied a special place among Chicago's neighborhoods, a destination for proud people reaping well-deserved rewards for years of hard work and deeply held values." This string of word choices—history, special, destination, proud, reaping, well-deserved, hard work, values—emphasizes the function of South Shore as a traditional first-house neighborhood in which working-class families move up to the lower middle class and beyond. "Now it's your turn to do the same," the brochure continues, and to do it in ways consistent with "the rich tradition of this storied community. LakeShore Pointe embodies everything you dreamed South Shore living could be."

What are the qualities that embody this dream? First, "friendly and secure." That is, you will have professional-managerial neighbors like yourself, and street crime is not as bad as people say it is. The booster story acknowledges and takes steps to defang the competing and well-known story of apocalyptic decline exemplified by "The Agony of South Shore." Second, LakeShore Pointe will be "comfortable and convenient." One powerful force driving the recent redevelopment and gentrification of the South Side lakefront has been its proximity to downtown—20 minutes by car, only a little more by bus or train, and the commuter train that runs on the I.C. tracks stops a block away.

Third, and crucially, living in LakeShore Pointe will be "ultimately fulfilling, for each and every homeowner as well as the community as a whole. It's a point of great pride for ShoreBank Development Corporation and its partners in our ongoing mission to serve as a catalyst for beneficial community redevelopment throughout the city's South and West Sides." The young, middle-class, black target market—figured in the brochure by a Romare Bearden-style collage of a bride and groom—will not have to choose between individual prosperity and a larger sense of belonging to a community, or "*The* Community," rooted in the Old Neighborhood.

The brochure urges those thinking of the suburbs to imagine instead that the upward-curving arc of their lives can be followed within the city, and it urges those who have moved to the suburbs to come back. Life remains more interesting here in the city, more convenient, cooler without being as dangerously edgy as South Shore was in the transition era of the 1960s, 1970s, and 1980s. And a profound satisfaction can be had in making your middle-class stand on the historic ground of black Chicago, the wellspring of The Community. The Irish, the Jews, they came through, got over, and moved on; you can follow the same path to individual success *and* hold the ground of the Old Neighborhood. Only in South Shore, where the good housing essential to middle-class security can be obtained relatively cheaply, will you be able to muster the resources to respond positively to *all* the claims of peoplehood that matter to you. Reinforcing that prospect is the South Shore Cultural Center, right across the street, which provides everything from jazz to day care. (Once upon a time, before it was the cultural center, that facility used to be the all-white, antisemitic South Shore Country Club. Then, after the neighborhood turned over and the country club expired, the Chicago police horse patrol took up residence in the club's disused stables and the Nation of Islam tried to buy the property, all before it became the cultural center.)

The brochure states that a condominium in LakeShore Pointe costs from $150,000 up; smaller town houses start at $180,000, the larger ones at up to $330,000: not much by the standards of Winnetka, but serious money on the South Side. By April 2000, thirty-one units had been sold, and a related development was in the works: a neo-Gothic apartment block across the street was being renovated into seventy-six condominiums.[16] The brochure, or at least the ideas it expressed, appeared to be convincing people to act. Or maybe they were just acting on the availability of relatively affordable upscale housing on the lakefront.

In any case, these developments were news. I would have been amazed if in the 1970s the South Shore Bank or anybody else had made such a substantial investment anywhere along 71st Street. I would have been amazed if anybody had built anything. I did not think of South Shore as a place where people built things. I thought of it as a place where people hung on—not just the preturnover residents who did not leave, but also the newcomers who tried to pick up the pieces of a neighborhood in which institutional order had been diminished and the informal networks of block life had been swept away by an overnight exodus.

If capital flight and the calculus of race hurt the neighborhood badly, the simple fact that South Shore's new residents often did not know one another (or the old residents) during the decades of transition made daily life especially difficult, and sometimes dangerous. "Nobody knew anybody then," Rhonda McMillon told me in 1997, remembering the 1980s, when she and her husband bought a house on the 7100 block of South Oglesby, a few doors down from the house I lived in as a small child. "There were kids running wild, and nobody knew whose kids they were, nobody would discipline them." If, a dozen years later, she was still "disappointed" in South Shore—the nightly gunfire in summertime, the drug dealers, the weak schools, the "Section Eight people" who "drop their trash everywhere"—she could at least feel the block-knitting effect of increasing familiarity. "Now I know all my neighbors, we're all connected, we help each other. We have a block watch, we call the police."

Growing up in South Shore as it turned over, I knew and in part lived by apocalyptic stories that explored the consequences of the sudden evacuation of neighborhood feeling from the neighborhood, the stories of the world's end that figure so prominently in the local mythos. The marketers who wrote the ShoreBank's brochure knew those stories, too. Rather than attempting an unconvincing refutation, they propose a triumphal third act that produces a counternarrative, turning decline and fall into transformation and rebirth. The brochure offers "South Side Living from a Different Point of View," which refers to both the new enclave's lake views and the revision of generic South Side stories that one must accept in order to invest in the development. In the play of narrative and counternarrative around the notion of the Old Neighborhood, the city of feeling and the city of fact take new shapes in relation to one another.

Dreams

The cover of the LakeShore Pointe brochure asks the reader to "*Stop. Look. Dream,*" which got me thinking about how to connect the "softest" neighborhood of all—the one we build in our dreams—to the historical facts of time and place.

In 1997 I went to see my family's first house in South Shore, the snug little three-bedroom on the 7100 block of South Oglesby that we sold in 1973 when we moved a few blocks over to a bigger house in the Highlands. When I went back to visit, our old neighbors were all long gone. On one side had been the Passmans, aging holdovers from Jewish South Shore. Were I in the business of dishing up a conventional Old Neighborhood soap opera, I would fill in the blanks of the obligatory nostalgic riff on my neighbors' ethnically redolent cooking with Mrs. Passman's blintzes.

On the other side lived the Thigpens. She was Canadian and white, he was American and black. They had a crimson-eyed, albino dog named Bigot, part German Shepherd and part shark, whose rage to sink her teeth into flesh seemed unnatural even for a dog. Bigot tore about the yard all day in a low-grade frenzy that escalated to high grade when anything moved within the acre she ambitiously imagined as biting range. When she barked furiously at me and threw herself slavering against the fence that divided the Thigpens' backyard from ours, Mrs. Thigpen would come out on the back porch, trailing cigarette smoke, and say "Bigot" in a low, flat voice that calmed the dog for a while. Bigot chased with cartoon-villain futility after a rabbit who lived in a thick tangle of bushes in the back of the Thigpens' yard, where a fence separated the yard from the alley. I refuse to call the thicket a briar patch, but it might as well have been. This rabbit was a veteran of urban life, a big mottled customer I saw only in pavement-colored blurs as it darted to and from sanctuary with the dog in pursuit. I have manufactured a fancifully detailed memory of the rabbit, though, as if it had stood, swell-chested, for a portrait: one-eyed, broken-eared, haunches pocked with bb shot, flanks scored by long transverse scars.

I cannot say that when I lived there I drew precise lessons from my situation between the Passmans and the Thigpens. I can say only that the two Thigpen boys I ran around with, Ricky and Alfred, had a certain touchy reserve that expressed to me—in concert with the mother's calm and the father's hard-working absence, the dog and the rabbit, the careful cordiality between the older Jewish couple next door and the more recently arrived black families all up and down the block—a delicate balance, a web of powerful forces arranged so that daily life on the block proceeded despite the tendency of those forces to seek destructive consummation.

When I came back there to visit in 1997, Emil and Carol, the family friends to whom we sold our first house in 1973, had moved away too. They were part of the last cohort of white families to buy a house in that section of South Shore. They had involved themselves deeply in neighborhood life for the best part of two decades—helping to run community organizations, pressuring landlords to repair or sell firetrap apartment buildings where drug dealers did business, buying another house on the block and fixing it up to rent out—but they had declared enough was enough soon after a kid from around the way tried to rob Emil at gunpoint outside a grocery store. Emil, a big man but gentle and slow to anger, surprised himself by grabbing the kid and wrestling the gun away from him. The kid took off running; the drug dealers from the corner, having gathered around to watch a white-haired Chicago Swede mix it up with one of their customers, were yelling "Shoot him! Shoot him!" The gun, though, turned out to be a heavy plastic fake, much to Emil's relief. The drug dealers, who apparently felt that amateur armed robberies were bad for business, gave the kid's name and address to the cops when they showed up, but no arrest was ever made. Emil and Carol bought an apartment with a grand lake view in a high-rise in Hyde Park.

Darryl and Tonia now owned the house. They were my age, in their midthirties, black, with big ambitions and two sound incomes. Although Darryl occasionally brooded over the big house in a different part of the South Side that got away, a Bronzeville mansion that could have been theirs if they had been able to raise more mortgage money, the house on Oglesby was still the fulfillment of a dream. Like many homeowners in South Shore, they had walked up and down the neighborhood's streets when they were younger, imagining what it would be like to own this house or that one. (The LakeShore Pointe brochure's "*Stop.* Look. *Dream*" refers to this common practice of fantasy first-house shopping in the neighborhood. Rhonda McMillon and her husband did it. So did Ron Grzywinski, one of the tight-knit group of new directors who took over the South Shore Bank in 1973. He told me he used to come to South Shore to look at houses when he was growing up in a mill neighborhood farther down the lakefront.) Darryl and Tonia did not have kids yet, but

South Shore's subpar schools would be a problem when and if they did. Like many of their neighbors, they planned to spend about 10 years in their house on Oglesby, their first, and then trade up. In the meantime, Darryl had gone into business on the side to get in on the redevelopment boom in Bronzeville, the Black Metropolis built by southern migration and subsequently gutted by urban renewal. He and a group of his friends had secured a small business loan from the South Shore Bank to open a market and deli a few miles away at 35th and King Drive, not far from the same housing projects whose imminent destruction would bring new neighbors—and fresh grist for the apocalyptic narrative mill—to South Shore.

Darryl, friendly and house-proud, showed me around and pointed out the refinished wood floors, the new deck that made the backyard seem even more cramped than it had been before, the conversion of the basement from dungeon to recreation room. Being in the basement dredged up a memory I had not thought about for a long time. In the late 1960s and early 1970s, when I was making my way through kindgarten and the early grades, I used to have a recurring nightmare in which a red fire alarm box and klaxon of the kind you see in old school buildings were mounted on the basement wall under the stairs in our house. In the dream, the alarm was ringing so loudly that I could not make out what the panicky announcer on the radio was shouting about. I suppose it was in part a cold war dream, fed by news of war and atomic competition, radio tests of the emergency early warning system, the air raid siren tests every Tuesday at noon, and the extended presence of uniformed soldiers in nearby Jackson Park. But I think it was also a South Shore dream, fed by panicky talk of urban crisis, by riots and the promise of more riots (the reason troops were deployed in the park), and by social upheaval proceeding so rapidly in the neighborhood that even small children felt it.

I should admit that my dream life makes other contributions, modest but lurid, to the tradition of South Shore apocalypticism. In addition to the fire alarm nightmare, I had the usual urban childhood anxiety dreams—like a recurring one in which I am in the kitchen late at night and somebody is rattling the locked door to the basement from the other side, or the one in which the bus home from school changes its route and misses my neighborhood and I am the only passenger left as it noses down increasingly alien and forbidding streets on its way to a depot far away on the West Side. But I also had what I think of as specifically South Shore dreams about assaults on our house. I used to dream about wild animals getting in: confronting a tiger bounding up the stairs, hearing roars and a four-footed commotion in the basement, finding the front door open and then sensing the stealthy movements of heavy bodies deep in the house. In my dreams I was forever trying and failing to seal the perimeter. Doors would not lock or would turn out to be too small to fill the doorway, windows would fly open for no reason, or I would get one door locked up and when I went to lock another, somebody would unthinkingly open the first one. In a typical dream, I would be hitting a lion on the head with a hammer on the narrow stairs up to the third floor, then rushing through the kitchen to help my brothers hold the back door shut while unseen forces tried to batter it down from outside.

About 1990, well after I moved away from the old neighborhood and soon after my parents sold our house and moved to Manhattan, my dreams of home defense in South Shore took a new turn. I still had to hold the house against outside threats, and that defense still forever failed to secure the infuriatingly porous perimeter, but for the first time the interlopers became people who got into the house to do unspecified bad things to my family and its property. Sometimes they were vague or half-seen figures; sometimes they were stringy-haired white guys displaying jailhouse affect; often, they were older and bigger versions of the black gangsters I used to avoid on 71st Street. All these characters descend, I think, from two thieves who broke into our house on Oglesby when I was about 7 years old and stole, among other things, a walkie-talkie set belonging to me and my brothers. The thieves were in and out before we got home, but the cops recognized "the m.o. of a salt-and-pepper team" (I remember trying to digest the new vocabulary) that was working the neighborhood. A few weeks after the break-in, I was playing ball in front of the house when a pair of hard-looking young men, one white and one black, came down the block, one on each sidewalk, cracking wise to one another on what I immediately recognized as my walkie-talkies. I never said a word to anybody about it. The lesson of the episode seemed obvious to me: people will routinely do bad things to you and get away with it, so why complain?

The interlopers in my dreams never quite did the bad things they were about to do, but there they were, barging through the front door, running up the stairs, struggling with me and my brothers on the second-floor landing. Sometimes these dreams began with me looking out the window from the second floor. Outside, a terrible, final momentum would be developing. A car had been abandoned slantwise and set afire in the middle of the block, a mob had gathered, somebody was down and getting a beating, people trooped out of the house across the street with loot in hand. Then I would hear a noise downstairs, rush to investigate, and the fighting would begin.

This is embarrassing mostly because it is so literal minded and crude. Once my family left the neighborhood, once South Shore stopped being an everyday social fact for me, my subconscious apparently felt free to stop fooling around with metaphors and to reduce the complexity of a satisfying life among good neighbors to the most overplayed clichés made readily available by our fearful public discourse about inner-city life. I was asleep when I dreamed these things, so I was not choosing these images and stories—I mean, if my neighbor commits mail fraud while I sleep, is *that* my fault too?—but I am a little scandalized that they can have a persistent life in me.

It does not matter, really, that my family suffered repeated burglaries, or that my brothers and I had to fend off semiregular attempted robberies and assaults in the street, or that home invaders really did shoot up our community-minded, God-fearing next-door neighbors on one spectacularly awful occasion. Nor does it much matter that I took my share of whacks, especially a memorable one to the head with a cutoff broom handle delivered circa 1974 by a kid wearing a pajama top (featuring a motif of horses' heads, I believe) who wanted my baseball glove. The fact remains that I lived a decent, mostly peaceable life in South Shore. Because I am impatient with people who equate my old neighborhood with uncontrolled violent crime, and especially because I devote some of my waking hours to writing and teaching in ways intended to break down conventions like "fear of crime" and get us to see them as, at least in part, historically situated bits of culture by which we know the messy world we live in, I am a little impatient with my subconscious for coming up with more of the usual stock mayhem. These dreams recur less frequently these days, and they have been largely supplanted by disoriented ones in which strange new upscale housing developments built by an advancing and invisible gentry have altered the neighborhood's landscape, but they do come back now and then.

So there I was in Darryl's basement. I verified, as I used to do when I was a kid, that there was indeed no fire alarm box on the wall under the stairs. Darryl and Tonia had finished and painted the walls and the concrete floor. They had put down some carpeting. There was a refrigerator and a television so Darryl could drink beer and watch football games with his buddies without messing up the living room, and he had a couple of weight benches set up for working out. The basement was so much smaller and brighter than I remembered it—so much less mysterious, so much less infused with terror—that I felt vaguely foolish for having used it to stage my subconscious dramas of household and neighborhood apocalypse.

The literature of South Shore addresses and expresses a deep pool of notions—some examined, some unexamined—that help to make up the inner lives of the neighborhood's citizens. Those inner lives play significant roles in real-world decisions about when to stay and when to go, who to vote for and what to buy, when and how much to commit to one's community, how far and how deep one's sense of that community extends. I am not saying that literature causes those decisions to happen (it does not), or that literature is important only to the extent, if any, that it affects such decisions (since literature for its own sake is good enough for me). Nor am I saying that South Shore somehow wrote its literature—since the words were written by individuals subject to all manner of creative influences other than that of the neighborhood. But I am saying that the literature of South Shore is a good place to look for models of inner life and its relation to the more measurable facts of neighborhood life, for maps of the cities of feeling that people carry around in their heads as they make their way through the city of fact. To consider how those inner lives might encourage or inhibit individuals' investment of themselves in the neighborhood is to consider the role of cities of feeling in the sustainability of cities of fact.

And vice versa. To know the serially troubled but ultimately consistent history of my old neighborhood is to begin to understand why that particular chunk of the city of fact seems to sustain certain cities of feeling and not others. Some stories recur persistently, chief among them the perennially popular narrative in which the world is about to come to an end. Looking back on my two decades in the neighborhood, I begin to see that it is even possible that apocalyptic narratives, a cultural tradition I had in common with my neighbors, actually functioned as cautionary tales on the order of Armageddon or Ragnarok. These stories reminded and guided us to do what we felt we could afford to do in the present—such as trying to behave decently toward one's neighbors, whoever they might be—to keep the peace in a delicately balanced local world seemingly always headed for dissolution in the near future. In ways that variously encouraged and discouraged investment in the neighborhood, these stories gave expression to the pervasive fear that our limited resources would not prove sufficient to the task of sustaining both individual well-being and the quality of neighborhood in the face of large-scale, often obscure, and therefore threatening urban processes.

Editors' Introduction to Chapter 5

When commentators like Evelyn Nieves of the New York Times *(2000) talk about devastated urban areas, they often point to East St. Louis, Illinois. Seen by the public imagination as thoroughly blighted, that city and its primarily African-American residents seem to have been thrown away during America's headlong rush toward economic growth and exurbia, treated as one of the trash heaps of America's urban history. Environmental justice scholar Robert Bullard (1994) again and again records the frequency and the virulence of that perception, which links African-American communities with throwaway communities. Turning that conventional imagery upside down, Kenneth Reardon has been reporting for a decade on an evolving collaborative partnership between the residents of East St. Louis's impoverished Emerson Park neighborhood and students and faculty at the University of Illinois at Urbana-Champaign. The evolving story of "participatory action research" appeared early as a practitioner's tale in an interview with John Forester and his students for the* American Sociologist *(Reardon et al., 1993). More recently, Reardon summarized the work of the East St. Louis Action Research Project for the First Person column of* Planning *(2000). This compelling tale can be understood as exemplifying Norman Krumholz and John Forester's "equity planning" (*Making Equity Planning Work, *1990). It can also be understood as part of a recent effort to have the urban (or regional) university "see itself as a citizen with responsibility to its neighbors," as David Maurrasse suggests in* Beyond the Campus *(2001). Told here with as much attention to what Rotella calls the city of feeling as to the city of fact, the details of Reardon's story offer the possibility for surprise in excess of any categorization.*

The terms and potential consequences of that less tangible investment can be found in the ways people act toward one another, the ways they imaginatively inhabit the landscape, the ways they think and talk and write about their neighborhoods, the stories they tell.

—Carlo Rotella

5

Ceola's Vision, Our Blessing: The Story of an Evolving Community–University Partnership in East St. Louis, Illinois

Kenneth M. Reardon

In the spring of 1987, Dr. Stanley O. Ikenberry, the former president of the University of Illinois at Urbana-Champaign (UIUC), met with members of the Higher Education Finance Committee of the Illinois State Legislature. During this meeting, state representative Wyvetter H. Younge (D-East St. Louis) challenged Dr. Ikenberry to demonstrate his commitment to urban public service by establishing an outreach project in East St. Louis. Dr. Ikenberry responded by reallocating $100,000 to launch the Urban Extension and Minority Access Project (UEMAP). Between 1988 and 1990, students participating in this initiative completed more than sixty projects investigating such local issues as waterfront development, stormwater management, downtown revitalization, housing rehabilitation, municipal reform, job generation, and environmental justice. However, student interest in UEMAP soon faded when local leaders showed little interest in these campus-generated plans and designs. In the spring of 1990, I joined UIUC's Department of Urban and Regional Planning as an assistant professor and was asked to assume responsibility for directing UEMAP.

Entering the Community

In July 1990, I made my first trip to East St. Louis, where the devastating impact of the city's long-term economic and fiscal decline was clearly evident. Two-thirds of the city's downtown office buildings and retail stores were vacant. All of the city's street lights and traffic signals were dark because of the municipality's inability to pay its electric bill. The city air smelled of burning garbage because of a 6-year hiatus in residential trash collection triggered by the city's fiscal problems. Forty percent of the city's building lots were vacant and 30 percent of its existing buildings were abandoned. I quickly realized that little in my previous 10 years of community organizing and urban planning practice in Paterson and Trenton, New Jersey; Hartford, Connecticut; and New York City had prepared me for the work I was about to undertake. Whereas the combined effects of suburbanization, deindustrialization, and disinvestment had destabilized select neighborhoods in these cities, East St. Louis appeared to be a city where most of its once-vibrant neighborhoods had been all but destroyed.

Shortly after arriving in East St. Louis, I began interviewing local leaders regarding their evaluation of UIUC's urban research. I soon learned that few neighborhood activists, church elders, social service directors, municipal officials, or area planners were aware of the university's 3-year-old community research project. I reframed my questions to elicit their interest in working together to establish a continuing community–university development partnership. One woman's response to this question reflected the opinion of many of the civic leaders I interviewed. "The last thing we need in East St. Louis," said Miss Ceola Davis, "is another university researcher who looks just like you telling us what any sixth grader already knows and having the gall to charge us $100,000 in state funds for the privilege."

The negative experiences of local residents who worked with university researchers during urban renewal, war on poverty, planned variations, and community development block grant programs had left them deeply skeptical of partnerships with higher educational institutions. From the residents' perspective, most urban researchers were little better than carpetbaggers, who used distressing census statistics to justify grants that provided significant benefits to their universities but few resources to the local community they studied. Many local leaders also believed that past research that had emphasized the serious problems confronting their city while ignoring their community's many strengths had helped create a climate in which few public or private-sector institutions were willing to invest in the local economy. Only one of the forty civic leaders I initially interviewed appeared willing to establish an ongoing partnership with the University of Illinois at Urbana-Champaign.

Bill Kreeb, the recently appointed executive director of the Lessie Bates Davis Neighborhood House, a 100-year-old settlement house, had come to believe that neighborhood residents and community-based institutions would have to become more involved in local economic development if conditions in the city's poorest neighborhoods were to improve. When I approached Bill regarding his assessment of local conditions and his agency's willingness to partner with the university, he appeared to be very interested. He told me how the neighborhood house had recently begun providing basic community organization and planning assistance to a small group of neighborhood residents who had formed an organization called the Emerson Park Development Corporation (EPDC) to improve the quality of life in the area immediately surrounding the neighborhood house.

Emerson Park was established in the early 1920s as a residential neighborhood composed of neat bungalows, garden apartments, and small apartment buildings for the Lithuanian, Ukrainian, and Polish workers who were then employed in the slaughterhouses, meat-packing plants, and railroad yards located in nearby National City. However, the long-term decline of regional meat-packing centers along with the nation's shift from rail-based to auto-based transportation subsequently eliminated most of the jobs held by local residents. The neighborhood's affluent families responded to these changes by moving, in the 1970s and 1980s, to areas with greater job opportunities. By 1990, the neighborhood's median family income had fallen to $6,738; more than 75.7 percent of its households were female-headed families with children; and the poverty rate was 54.9 percent. In addition, more than 68.7 percent of the neighborhood's building lots were vacant and more than 38 percent of its buildings were abandoned following decades of outmigration.

Bill Kreeb offered to introduce me to the members of the Emerson Park Development Corporation, which was headed by Miss Davis, a long-time community worker employed by the neighborhood house. A week later, at a meeting arranged by Mr. Kreeb, she described how she and her neighbors had established the Emerson Park Development Corporation. Most of the women I was meeting, she said, had children or grandchildren who attended the Lessie Bates Davis Day Care Program. More than a dozen years earlier, these mothers and grandmothers had become increasingly alarmed by the physical deterioration and illegal drug sales taking place within the neighborhood. Since the city and county appeared unwilling to take action to protect their children, families, and homes, the women decided to launch their own community revitalization initiative.

A small cadre of older neighborhood women initiated this process by recruiting unemployed railroad workers to convert three abandoned buildings facing the Lessie Bates Davis Day Care Center into a toddlers' playground. She herself had led a neighborhood delegation on a day-long bus excursion to the county administration building located in suburban Belleville to determine who owned these derelict structures. In Belleville, the group discovered to their surprise that the trustees of St. Clair County had acquired title to these properties when their former owners had failed to pay their local property taxes.

She subsequently organized a much larger group of residents to attend a public hearing of the St. Clair County Property Tax Disposition Committee to request site control of these tax-foreclosed properties. The committee agreed. Within weeks a group of local volunteers was disassembling the delinquent structures brick by brick. In the process, residents carefully saved every salvageable item, including bricks, fixtures, woodwork, wiring, and pipes. With the help of a local black contractor, they hauled these materials to a nearby salvage yard from which they received $3,000. They raised an additional $1,000 for the playground project by selling fried chicken and catfish dinners on Friday nights. With $4,000 in hand, EPDC approached the corporate philanthropy department of Ralston-Purina for matching funds. While Ralston, a large, multinational corporation with a history of socially responsible philanthropy, liked the playground project, it was only willing to match EPDC's investment 25 cents on the dollar, given the group's modest track record.

With the assistance of dozens of neighborhood volunteers, local contractors, and settlement house workers, Miss Davis directed the construction of the playground called Shugue Park after EPDC's first president. This 5,000-square-foot green space featured a well-maintained lawn, numerous planting beds, a sitting area, water fountain, and a modest sign. Thirteen years after its construction, Shugue Park remains a source of great pride and inspiration for local residents. The success of this project galvanized EPDC's desire to rebuild Emerson Park through a series of large-scale development projects.

When Miss Davis finished her story of EPDC's park project, she asked me to describe the kind of work my students and I were prepared to do in East St. Louis. I told her that we were trained to work with community groups, municipal agencies, and private firms to develop comprehensive economic, social, and physical development plans designed to improve the quality of life in distressed urban and rural communities. She then asked me to describe one or two of the projects that my students and I had completed. I explained how, in the face of powerful redevelopment forces, we had recently helped the Essex Street Merchants Association in New York City save a cooperatively managed public food market serving the city's newest immigrants. Miss Davis then asked me if I had ever worked in an African-American community. I told her of my previous work as a community organizer in North Hartford, Connecticut, and as an urban researcher in the Bedford-Stuyvesant section of Brooklyn. She concluded our meeting by asking me to send her group copies of my Essex Street Market Plan and Bedford-Stuyvesant Housing Report as well as references from the grassroots leaders with whom I had worked in these neighborhoods.

With the fall semester rapidly approaching, I felt increasing pressure to identify an East St. Louis project that my students and I could undertake. Several days later, I gratefully received an invitation from Miss Davis to a second EPDC meeting. Miss Davis, her group, and five Baptist ministers from the area greeted me. Following friendly introductions, the Reverend Robert Jones, president of the Metro East Area Project Board, a faith-based organizing project affiliated with EPDC, asked me why the University of Illinois had come to their community. He asked, "How long is the university committed to working with our city's neighborhoods? Is the university willing to struggle against the machine politics of East St. Louis to accomplish its work? Who does the university see as their primary partners in the city—the people, its churches, local government or state representative Younge? What experiences do you have working with African-American communities? How long is your personal commitment to this project? Will the university reassign you if local political leaders challenge your work?"

After I responded to Reverend Jones's questions, Miss Davis said that her group would enter into a partnership with my department provided we agreed to five basic principles of community–university engagement that she and her neighbors had developed. First, the residents of Emerson Park and their organization, EPDC, would determine the local issues that the University of Illinois would work on. Second, local residents would have to be actively involved with UIUC students at every step of the research and planning process. Third, UIUC's Department of Urban and Regional Planning must make a minimum 5-year commitment to Emerson Park following its 1-year probationary period. Fourth, the university must help EPDC gain access to regional funding agencies to secure the resources needed to implement local development projects. Fifth, the university must help EPDC create a nonprofit organization to sustain the community revitalization process after the university left the community. Miss Davis then thanked me for attending the meeting and encouraged me to discuss EPDC's partnership terms with my colleagues; Reverend Jones offered me "traveling mercies" for my 200-mile return trip to Champaign-Urbana.

Although I was moved by the "traveling mercies" offered me by those whose daily lives I viewed as more hazardous than my travel, I felt deeply conflicted during my return trip to the UIUC campus. I wanted to work with Miss Davis's group yet regretted their lack of enthusiasm for our initial offer of assistance. It would take years for me to fully understand the residents' caution regarding "free gifts" from white outsiders representing powerful institutions such as the University of Illinois. The following morning, I told Professor Hopkins, then head of UIUC's Department of Urban and Regional Planning, "There is good news and bad news regarding the future of our East St. Louis project." The good news was the marvelous group of residents from the city's poorest neighborhood, who were supported by local faith-based organizations and social service agencies and absolutely committed to turning their community around. "What's the question?" he responded. "We should definitely work with these people." I handed him a typed version of Miss Davis's five principles for successful community and university partnerships. Smiling, he told me to tell Miss Davis that we would welcome the opportunity to work with her group and do all that we could to honor the spirit and substance of her principles.

I called Miss Davis to tell her we welcomed her group's invitation to work in Emerson Park under the terms she had presented. Several days later, I returned to Emerson Park with the suggestion that we begin our collaborative effort by working with residents to create a 5-year stabilization plan to address the most serious problems facing the neighborhood. Miss Davis and her neighbors objected; they had been "studied to death." They argued, instead, for the immediate implementation of a second, larger, physical improvement project similar to Shugue Park. Among the specific projects they advocated were the transformation of a local motel used for prostitution into a legitimate business, rehabilitation of the area's Craftsman bungalows, and reclamation of the neighborhood's City Beautiful-Era Exchange Boulevard.

I asked them if they would, as members of the city council or county board, recommend funding for such ambitious projects in the absence of a blueprint revealing how these initiatives might contribute to the area's short-term stabilization and long-term revitalization. I reminded them of the intense competition that existed for discretionary housing and community development funds. They discussed my counterargument and agreed, somewhat reluctantly, to work with my students on this effort provided it led to the implementation of specific neighborhood improvement projects within the year. We further agreed to a joint program of work for the 1990–1991 academic year that featured a semester of planning followed by a semester of neighborhood improvement projects.

I recommended the use of participatory action research methods to involve local residents as "coplanners" with university students at every step of the planning process. I also suggested that we change the name of our project from the Urban Extension and Minority Access Project (UEMAP) to the East St. Louis Action Research Project (ESLARP) to emphasize this new collaborative approach. I described how this "bottom-up, bottom-sideways" model had been successfully used by teams of local activists and university researchers around the world to address the unintended negative consequences of globalization, including flight of capital. I then described how participatory action research integrated the local knowledge of community residents with the expert knowledge of professionals to produce innovative solutions to messy social problems while simultaneously enhancing the organizational capacity of community-based institutions. I finished my remarks by explaining how residents of Chicago's poorest neighborhoods had used this approach to create highly effective neighborhood-based planning organizations that eventually coalesced with the city's civil rights, community development, and good government groups and with trade unions to elect Mayor Harold Washington, whose administration championed many of the economic and community development policies they sought.

Creating the Emerson Park Plan

Nine graduate and two undergraduate students signed up for my neighborhood planning workshop in the fall of 1990. During the first class I informed the students that the Department of Urban and Regional Planning had agreed to assist the residents of East St. Louis's poorest neighborhood in creating a 5-year comprehensive stabilization plan to halt the exodus of businesses and residents from the area. They would be working with leaders of a newly formed community development corporation (CDC) to create a plan to improve the overall quality of life in East St. Louis's Emerson Park neighborhood. In the days following this first class, two members of the workshop assisted me in devising a preliminary proposal for a participatory action research-based approach to creating a community stabilization plan for Emerson Park.

During the second week of the semester, the entire class traveled with me to East St. Louis to gain a better understanding of the city's physical structure and social history. My students chatted excitedly during the 3-hour trip; however, as our van approached the city, they grew very quiet. I watched them as they gazed at the city's blocks of abandoned industrial buildings, railroad yards, commercial areas, and single-family homes. By the time we arrived at the Lessie Bates Davis Neighborhood House, the van was dead silent. As the van stopped in front of the neighborhood house, four neighborhood leaders whom we would come to know very well warmly greeted us—Miss Davis, Mr. Richard Suttle, Miss Peggy Haney, and Miss Kathy Tucker. Inside, we met with fifteen residents to identify the major issues confronting their neighborhood, review the basic work plan we had developed, and scrutinize the letter of understanding we had drafted outlining the roles each party would play in the planning process. The residents took nearly an hour to describe the serious unemployment, housing, health care, transportation, education, public safety, industrial pollution, and youth development problems confronting their area.

We then described how we planned to involve local residents and institutional leaders in collecting and analyzing the data needed to document these concerns and in formulating innovative solutions to these problems. In doing so, we emphasized how participatory action research methods would be used to actively involve local residents as coinvestigators with university-trained researchers at every step of the planning process. While the majority of those attending the meeting appeared eager to participate in this bottom-up, bottom-sideways planning process, one woman asked, "Why do we need the university if we are going to do all of the work?" Miss Davis responded to her question by emphasizing the importance of training local residents to do their own research. The residents proposed a monthly planning meeting so that university personnel and neighborhood residents could together analyze the data we would be collecting.

Three weeks later two of my students and I were back at Emerson Park to distribute flyers announcing our first community planning meeting. With the assistance of several EPDC leaders, we knocked on every door in the neighborhood, encouraging residents to participate in the upcoming meeting. Thirty residents attended this meeting at which my students presented a subset of the twenty-nine tables containing population, employment, income, and housing data from the 1960, 1970, and 1980 censuses which they had developed. Near the end of the students' presentation, a young man stood up and addressed the students. After thanking them for their good work, he asked them why they had limited their comparison of the neighborhood's demographic trends to those of the city and the county. Why had they not included a comparison with the suburban ring communities surrounding East St. Louis, which he felt would have highlighted the increasingly uneven pattern of development characterizing the Greater St. Louis Metropolitan Region? As the student presenters listened to this young man's thoughtful critique, they nervously looked to me to provide a defensible explanation for our choice of comparison groups. I thanked him for his contribution, agreed with his criticism of our approach, and committed our students to revising our demographic analysis of the neighborhood to include the additional comparisons he suggested.

On our return trip to Champaign, the students complained about the additional work required to change their census tables, suggesting this might be a good job for a new assistant professor. But I reminded them that such reanalysis of data was common in the early phases of most research projects. It was arrogant to think that only university-trained planners could have insights about how to best understand the data for a particular study area. Such give and take with community residents about the best research design, survey instruments, and data analysis was, in fact, the hallmark of the participatory action research approach that we had proposed in response to Miss Davis's principles for community partnerships. In the days following this exchange, the class developed two instruments to be used in conducting a land use, building conditions, and site maintenance survey of Emerson Park's building parcels and a sidewalk, curb, street, and sewer survey to be used in evaluating the municipal infrastructure serving the neighborhood.

Using reformatted student evaluation forms that were machine readable, twenty students were able to quickly collect data on twenty-two aspects of each of the neighborhood's 1,407 building lots and sixteen aspects of each of the neighborhood's 220 block-long street lengths. The use of these machine-readable data forms enabled our class to return to Emerson Park within a week of our initial data collection effort to share information on existing physical conditions using nine wall-sized maps and a packet of tables (see figures 5.1 and 5.2). Before we presented these data to the forty residents attending our second community planning meeting, we took time to review several of the most important demographic charts that we had changed in response to the criticisms raised by the young man at our first meeting. As the students explained the impact of this new comparison on our analysis, the young man smiled at several of his neighbors, who gave him a nod in recognition of his contribution to our analytical effort. In the years following this event, several experienced leaders told me what a profound effect our willingness to accept a neighborhood youth's criticism of our work had upon their desire to work with us.

Having completed our analysis of the neighborhood's demographic profile and physical conditions, our class began working with EPDC's executive committee to design two interview schedules. We used the first

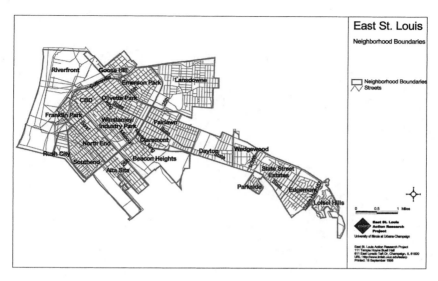

Figure 5.1
East St. Louis neighborhood boundaries. (Source: East St. Louis Action Research Project.)

interview schedule to elicit residents' perceptions of existing neighborhood conditions, visions for the future, and specific improvement proposals to promote neighborhood revitalization. We designed the second schedule to elicit similar information from the leaders of such institutions as schools, churches, businesses, and social service agencies located within the neighborhood and other institutions located outside of the Emerson Park study area but responsible for providing services to its people.

During the month of November, with the assistance of volunteers recruited from planning students and EPDC executive committee members, the class was able to knock on the doors of Emerson Park's 900 households and interview a total of 140 individuals. Student and community researchers also interviewed leaders of more than forty nonprofit and government agencies serving the neighborhood. The participation of residents in the interviewing process prompted the overwhelming majority of those whose doors we knocked on to invite us into their homes. This participation also helped our students more accurately interpret the comments made by local residents.

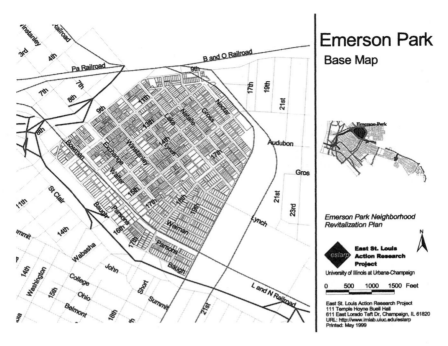

Figure 5.2
Emerson Park neighborhood base map. (Source: East St. Louis Action Research Project.)

These interviews were a turning point for many of the students and EPDC executive committee members. Many students and EPDC leaders believed, for example, that the majority of neighborhood youths did not care about the area and were in part responsible for its decline. The experiences that our students and their community partners had visiting the homes of local residents dispelled this and many other unexamined stereotypes of the neighborhood's residents. The stories that residents told of their struggles to purchase homes in Emerson Park and their subsequent efforts to maintain their families and properties in the face of the city's deepening economic and fiscal collapse greatly affected our students. They listened to young mothers describe their efforts to support their families while working at several low-paying jobs in areas poorly served by mass transit. They heard retirees who had worked for companies with no pension programs describe the difficulties of making ends meet in the face of escalating utility bills, property taxes, and prescription drug costs. They learned of the fear small children experienced traveling to school on streets dominated by drug dealers, addicts, paid lookouts, and streetwalkers.

What our students learned during these interviews prompted them to reexamine what they thought they knew about our country, the planning profession, and their own futures. The interviews helped them realize how closely the hopes, aspirations, and dreams of Emerson Park's residents reflected those of their own families and how economic conditions and social policies had greatly complicated the futures of many East St. Louis residents while advancing their own. My students' reflections upon these realities, along with the growing number of caring relationships they were developing with local residents, solidified their commitment to the project. As the semester drew to a close, the students began scheduling additional trips to Emerson Park to collect data they felt were needed for their plan. In early December, the whole class traveled to Emerson Park to do one final round of door-knocking in preparation for a neighborhood-wide meeting scheduled by EPDC to review the first draft of the Emerson Park Neighborhood Improvement Plan.

More than 140 neighborhood residents and elected officials attended this community meeting to review the major findings and recommendations contained in the preliminary draft of the plan. During the first half of the meeting, my students presented the results of their team's demographic analyses, physical condition surveys, resident interviews, and institutional interviews. During the comment period that followed this presentation, several long-time residents congratulated the team on the quality of their work. Others gently chided them for failing to address school enrollment and academic performance and also failing to address the absence of local church involvement in neighborhood organizing, planning, and development efforts.

During the second part of the meeting, several EPDC leaders presented the plan's overall goal, development objectives, and specific neighborhood improvement proposals. The primary goal of the plan was to stabilize Emerson Park as a mixed-use, mixed-income neighborhood attractive to a wide range of East St. Louis residents through the implementation of an integrated community development strategy featuring neighborhood beautification, housing development, substance abuse prevention, economic development, and community organizing initiatives. The plan concluded with a Betty Crocker Cookbook-like set of instructions for carrying out a modest set of community organizing, public safety, and neighborhood beautification projects during the next year. These projects were presented in a developmental fashion, beginning with modest efforts that could be implemented with a few volunteers using minimal funds and progressing to larger-scale projects requiring significant numbers of volunteers, some paid staff, and higher levels of external funding. Several "Amens" were voiced and many heads nodded as EPDC's leaders presented the basic structure and specific proposals for the coming year. When the leaders concluded their presentation, Dr. Kenneth Bonner, a long-time Lessie Bates Davis board member, addressed the audience, "Our Illinois students have done an excellent job giving voice to our collective aspirations for our neighborhood and our city. It is now up to each and every one of us to put our shoulders to the wheel to turn this prose into reality. Working together, I know we can do this."

Implementing the Emerson Park Plan

Elated by the residents' response to our first community planning effort, the students asked me how I was planning to follow up on their work during the coming semester. I was not sure, given my spring teaching duties, which included two undergraduate lecture classes. Several of the students proposed to speak with the chair of our department about the necessity to maintain the Emerson Park project's momentum via a continuous workshop. Convinced by the students, Professor Hopkins, the chair, asked me the next day if I would be willing to lead an "encore" performance of my neighborhood planning workshop during the upcoming semester in lieu of the undergraduate lecture courses I was scheduled to teach. In the weeks that followed, the eleven students who had participated in the fall workshop recruited eighteen of their fellow students to continue the Emerson Park project.

During the intersession, forty-five neighborhood residents and I participated in a meeting, on Martin Luther King's birthday, that examined alternative strategies for implementing the Emerson Park Neighborhood

Improvement Plan. Following considerable discussion, the group decided to concentrate on three major activities during the coming year. First, they committed themselves to working together to transform the Emerson Park Development Corporation into a 501c3 organization authorized to receive and distribute tax-deductible donations for neighborhood development. Second, they decided to raise the funds needed to complete several neighborhood beautification projects to convince skeptical residents and officials of EPDC's effectiveness. Finally, they asked me if our students would work with them to devise a community-based substance abuse and crime prevention program to address the neighborhood's growing substance abuse problem.

I began the spring workshop with a panel presentation by students from the fall class who laid out the key elements of their Emerson Park Neighborhood Improvement Plan for the new students. The class then decided to divide itself into subgroups to develop strategies for working on EPDC's incorporation initiative, neighborhood beautification campaign, and substance abuse—violent crime prevention program. Within 3 weeks, the incorporation subgroup was working with EPDC's leaders and Land of Lincoln Legal Services attorneys to develop a constitution, bylaws, and incorporation papers for the organization. The neighborhood beautification subgroup was busy assisting EPDC in identifying an area where they could launch their environmental cleanup efforts and in recruiting volunteers to participate in a weekend cleanup of illegally dumped trash. Finally, the crime prevention subgroup had organized a data-gathering trip to the National Center for Crime Prevention in Nashville, Tennessee, to collect information on model crime prevention programs.

As we approached the midpoint of the spring semester, we were notified by the American Institute of Certified Planners that the Emerson Park Neighborhood Improvement Plan had been awarded the "Best Project Plan Award" for 1991. Resident and student enthusiasm for our newly established community development partnership soared as members of the fall neighborhood planning workshop traveled to Washington, D.C., to collect plaques and certificates in recognition of their plan.

However, our excitement was quickly tested when more than a dozen local community development funding organizations refused to support even the least costly elements of our initial plan. These funders believed the long-term deterioration of East St. Louis's economy made any commitment of new funds a highly risky proposition. I dreaded EPDC's March membership meeting when I would have to tell the neighborhood women who had accomplished so much in the face of such great opposition that not one of the major funders in the Greater Metro East Area was willing to support any of the proposals contained in their nationally recognized plan. When, at the end of that meeting, I nervously informed the group that we had failed to raise a single dollar to implement the Emerson Park Neighborhood Improvement Plan, Miss Davis sensed my embarrassment. She put her arms around me and said, "What did you expect—a check?" Imploring the group to use the "resources at hand," Miss Davis asked those attending the meeting to look around the room. "We have sixty residents and twenty students who have worked together for nearly a year to create a plan for transforming our neighborhood. We must remember that collectively we have resources! There are young people in our families who can clear lots, we have tools in our cellars and garages that can be used to cut weeds, shrubs and trees, and some of us own pick-ups that can be used to transport illegally dumped trash to the landfill." She concluded her remarks by suggesting that we organize a cleanup of 9th Street, the site of the neighborhood's worst illegal dumping, drug sales, and prostitution.

Neighborhood residents, local institutions, and my students responded energetically to Miss Davis's call to arms! Within 2 weeks, EPDC organized dozens of neighborhood residents to clear lots, operate heavy equipment, and feed volunteers. My class organized students from the Department of Urban and Regional Planning, the Department of Chemistry, Hillel, the Tri-Delta Sorority, the Muslim Students Association, and St. John's Lutheran Church Boy Scouts to participate in the cleanup. At 6 a.m. on the day of this event, we boarded three yellow school buses for the trip to East St. Louis. Professor Lew Hopkins and I assisted Miss Davis in organizing mixed groups of students and residents to remove trash from 9th Street's ten most trash-filled lots. As we began our work, state representative Younge drove up in a station wagon wearing dungarees, a work shirt, and gloves.

Volunteers filled increasing numbers of plastic bags with trash. However, without funds to transport the trash to the nearby landfill, we had no place to put the collected garbage. So Miss Davis and Mr. Suttle instructed the students to line the trash bags up in neat piles on both sides of 9th Street. They were certain that this garbage would be quickly removed by St. Clair County's Public Works Department, whose managers would not want the appearance of this busy county road to be compromised by stacks of trash. At the end of the cleanup, more than 50 yards of 9th Street were neatly lined with 5-foot-high walls of carefully bagged garbage. Soon after we finished stacking the last of the trash bags, a television crew from the local NBC affiliate in St. Louis unexpectedly appeared. As the reporter began to film the garbage along 9th Street and interview the local residents and our students, I noticed Miss Davis and Mr. Suttle smiling in the background. When I asked them if they had anything to do with the timely appearance of the reporter, they just chuckled.

EPDC and the university received considerable positive attention from the press following this successful cleanup effort. In addition, the Lessie Bates Davis Neighborhood House received a $15,000 grant from a local community foundation to assist EPDC with their neighborhood beautification efforts. At the community meeting following this announcement, I expected EPDC to decide to use these funds to hire local unemployed youths to continue the cleanup program. However, the residents concluded that they would never be able to clean up all of the neighborhood lots covered with illegally dumped trash if they paid themselves to do so. Instead, they decided to continue the cleanup effort on a voluntary basis, using the grant funds to rent the heavy equipment needed to transport the trash to a nearby landfill and to pay the tipping fees. During the summer and fall of 1991, EPDC, with the assistance of the Lessie Bates Davis house, Catholic Urban Programs, and the university cleared several hundred lots and alleys of trash.

The success of EPDC's cleanup campaign generated new interest in the organization's public safety report, which my students completed in May 1991. By the fall of 1991, representatives of the East St. Louis Public Housing Authority agreed to demolish several long-abandoned public housing projects located in the northernmost portion of the neighborhood that were "hot spots" for various illegal activities. One such building had earned the moniker "the Pharmacy," given its role as a major distribution point for the sale of illegal drugs. In addition, the housing authority announced plans to rehabilitate approximately forty town house units located near Lessie Bates Davis that we had recommended in our plan.

Several weeks after this meeting, representatives of the U.S. Attorney for the Southern District of Illinois met with EPDC's leaders to announce an ambitious undercover initiative aimed at dramatically reducing the neighborhood's growing crack-cocaine problem. Within 3 months of this meeting with Mr. Bruce Reppert, assistant U.S. attorney for the Southern District, a series of drug raids were held in Emerson Park, which removed the network of street dealers who had taken over many sections of the neighborhood. The intensity of this federal enforcement effort, in combination with a community policing program launched by the city, significantly reduced the level of illegal activity within the neighborhood.

Encouraged by the success of their neighborhood beautification and crime prevention efforts, EPDC requested university assistance for a new housing improvement program. Local leaders believed they could attract public and private housing investment to Emerson Park if they could demonstrate its residents' commitment to maintaining their homes, streets, and neighborhood. Students enrolled in my third neighborhood planning workshop, assisted by others participating in Michael Andreja-sich's housing design studio and Brian Orland's senior landscape design studio, worked with EPDC's executive committee to identify single-family homes owned by low-income individuals or couples that required minor exterior painting and repairs. Architecture students subsequently worked with EPDC leaders to complete "scope of services" reports describing the work to be completed at each residence, while planning and landscape architecture students approached area lumberyards, paint stores, home improvement centers, and nurseries for donations. In October 1991, forty students, accompanied by Mike, Brian, and myself, traveled to Emerson Park to complete modest exterior home repairs that included patching roofs, repointing brick porches, installing guttering, and painting homes owned by the neighborhood's poorest residents. In many cases, our students were trained to complete this work by African American contractors living in the neighborhood. The dramatic impact that the improvement of four adjacent houses had on the neighborhood's physical appearance produced requests for assistance from many other neighborhood residents.

Figure 5.3
Student volunteers working on the construction of a new home as part of EPDC's faith-based family housing program. (Photograph by R. Selby, ESLARP.)

The success of this joint EPDC–UIUC housing venture resulted in it becoming a continuing program. During the past 9 years, more than 3,000 UIUC students have traveled to East St. Louis for "work weekends" to clear trash from abandoned lots, repair and paint the homes of low-income residents, renovate schools and community centers, build new housing, construct playgrounds and parks, and erect a farmers' market (see figures 5.3, 5.4, and 5.5). In the spring of 1992, the Illinois state treasurer, who had been watching the progress of EPDC-UIUC's self-help housing initiative, called to offer financial assistance. Peter G. Fitzgerald, then Illinois state treasurer and currently U.S. senator, offered to deposit $5 million in state tax receipts in a local bank so the difference in interest rates demanded by the state and those offered by the banks could be used by EPDC to establish a low-interest home improvement loan program. This 3-year commitment by the state treasurer helped capitalize a $75,000 low-interest loan fund that enabled neighborhood residents whose homes required more than an exterior power washing and painting to purchase needed building materials. Working under the joint

Figure 5.4
Youthful volunteers participating in the construction of the Illinois Avenue playground. (Photograph by K. Reardon.)

Figure 5.5
Opening day of the East St. Louis farmers' market.

supervision of local building trades professionals and architecture faculty, UIUC students could use these materials to complete larger-scale projects, such as rebuilding porches, replacing exterior staircases, and installing new roofs.

Creating a Sustainable Community Development Corporation

In early 1993, as work proceeded on the three objectives identified by EPDC's Executive Committees in the spring of 1991, a group of inner-city and suburban churches with ties to Lessie Bates Davis Neighborhood House and Catholic Urban Programs approached the organization with a proposal to develop a faith-based family housing program to construct affordable homes on several nearby vacant lots. With the help of UIUC students, faculty, and alumni, plans were developed for four 1,200-square-foot homes to be built in Emerson Park and the nearby Olivette Park neighborhood. In the spring of 1995, four homes designed by UIUC architecture graduate Don Johnson were built in these neighborhoods with the assistance of dozens of UIUC students. Four additional homes subsequently built by the same coalition prompted the City of East St. Louis's Community Development Block Grant Office to invite EPDC to apply for $189,000 in HOME funds from the U.S. Department of Housing and Urban Development (HUD). With the university's assistance, EPDC completed major renovations for seven low-income homeowners. In 1997 it hired a graduate of UIUC's East St. Louis Action Research Project, Ms. Vickie Forby, to assist the organization in expanding its membership base, strengthening its leadership core, increasing its budget, and diversifying its funding base.

Shortly after Ms. Forby became EPDC's first staff member, I received a call from Miss Davis, who had learned that the East-West Gateway Coordinating Council, the region's transportation agency, was preparing plans to extend St. Louis's highly successful light rail line from downtown St. Louis across the river to Scott Air Force Base in southwestern Illinois. Miss Davis was eager to see the agency use an abandoned, state-owned, railroad right-of-way that ran through East St. Louis's poorest residential neighborhoods, including Emerson Park, as the route for the proposed light rail line. Miss Davis believed dozens of Emerson Park residents would be able to gain access to living-wage jobs in St. Louis's central business and international airport districts if this alternate route was chosen and a train station could be built in the neighborhood. Miss Davis asked me if I could recruit a group of planning and design colleagues to prepare a report highlighting the environmental, economic, and social benefits of using the existing right-of-way. She also asked for our assistance in securing funds to purchase land around our proposed rail stations so EPDC and its neighboring community organizations could be in a stronger position to bargain with would-be investors interested in what could soon become one of the region's hottest real estate corridors.

In the fall of 1997, Professor Robert Selby from the university's School of Architecture and Professor Orland from the Department of Landscape Architecture began working with approximately forty design students to complete the light rail routing plan requested by Miss Davis. As they worked on this initiative, Don Johnson, one of our former students who has emerged as East St. Louis's most successful developer, entered into negotiations with the Southwestern Illinois Development Authority to secure funds for land assembly that would permit EPDC to acquire a strategic number of parcels surrounding the proposed Emerson Park rail station site contained in our plan. As the holidays approached, Miss Davis assisted EPDC in organizing a neighborhood meeting to discuss the proposed routing for the light rail line extension.

As a result of Miss Davis's organizing efforts, nearly a hundred local residents, area officials, and newspaper reporters attended the light rail meeting held at the neighborhood house. Miss Davis began the meeting by thanking East St. Louis Mayor Bush for attending and explained the deep interest Emerson Park residents had in the safety, economic development, transportation, and aesthetic dimensions of the light rail routing proposal. Miss Davis then went on to describe the efforts local residents had made, with the help of the university, to identify the optimal route for the rail extension. With these remarks, she introduced professors Selby and Orland, whose students presented the case for the community residents' preferred route. When the students finished, the mayor asked the residents to indicate their support for Miss Davis's plan by raising their hands. The mayor and Miss Davis smiled as all of the children, adults, and students attending the meeting raised both of their hands. Turning to the local television and cable station cameras covering the event, the mayor complimented the students on their work, voiced his strong support for Miss Davis's proposed route, and invited those in attendance to come to the next city council meeting, where he promised to introduce a motion in support of their routing plan.

More than seventy-five Emerson Park residents crowded into the East St. Louis City Council chambers the next week to watch the mayor and the council unanimously approve EPDC's proposed rail route. Neighborhood leaders smiled as the council voted in favor of their route, denying the councilors' usual political allies the financial benefits of having their property purchased at inflated prices for a route that would be more costly and could endanger the children attending schools and participating in after-school programs sponsored by institutions located adjacent to the original route. Within a month, the East St. Louis Gateway Coordinating Council held hearings that culminated in the approval of EPDC's routing plan.

In the spring of 1998, following a presentation I gave at St. Louis University on the improved investment climate in East St. Louis, Richard Baron, a distinguished older gentleman, approached Vickie Forby, Richard Suttle, and me to introduce himself. He was answering my call for a "few good men and women" to bring to life a major mixed-use, mixed-income development in the area surrounding the neighborhood's planned rail station. "I have been wanting to work in East St. Louis for many years," he said. "I would welcome the opportunity to come to your office to speak with you about what my firm has been able to do here in St. Louis, in Kansas City, and in Chicago to create the kind of projects you described in your presentation."

Within a week, Mr. Baron visited EPDC's offices at the neighborhood house to discuss a $30 million affordable housing project that he wanted to build on the land that EPDC owned adjacent to Emerson Park's proposed rail site. By the end of the month, Mr. Baron had met with EPDC's executive committee several times to discuss the specifics of his proposal. Following these interactions, EPDC decided to enter into a development agreement with Mr. Baron's firm to create 225 units of rental housing and 70 units of new single-family housing. The Parsons Place Project, as conceived by Mr. Baron and EPDC, will feature 60 percent market rate housing and 40 percent subsidized housing financed, in part, by Low-Income Housing Tax Credit funds provided by the Illinois Housing Development Agency. Under the agreement, Mr. Baron's firm will share a portion of its developer's fees with EPDC and will allow the organization to designate the local builder to construct the project's single-family units. By the spring of 1999, Mr. Baron and EPDC had secured all but $7 million needed for the project. In the fall of 1999, the City of East St. Louis completed the sewer, street, sidewalk, curb, and lighting improvements required by the project.

Meanwhile, as early as 1995, the university had begun to receive an increasing number of planning requests from nearby neighborhoods and communities whose residents and officials had observed the dramatic changes taking place in Emerson Park. UIUC was able to respond favorably to many of these requests because of the university's willingness to increase its support for the project from $100,000 to $200,000 and HUD's decision to award a $500,000 Community Outreach Partnership Center Grant to UIUC. This grant, supplemented by university and city funds, in 1996 enabled UIUC to establish a permanent community planning and design assistance office in East St. Louis called the East St. Louis Neighborhood Technical Assistance Center (NTAC). During the past 5 years, NTAC's architect, urban planner, and nonprofit management specialist have assisted more than sixty community-based organizations, faith-based institutions, and municipal agencies in planning and designing more than a hundred neighborhood improvement projects in the Greater East St. Louis area.

Epiphany in Effingham

As the pace of resident-led economic and community development began to quicken in Emerson Park and its surrounding neighborhoods in late 1995, my colleagues and I were invited to what we have come to refer to as "the come to Jesus meeting." At the end of a meeting nominally called to discuss our future staffing plans, Miss Davis asked us if we still believed in the "empowerment planning model" we had spoken about when we first came to town. I confidently stated that we most certainly did. "When do you plan to start doing this kind of work in East St. Louis?" she asked. Seeing that I was taken aback, Miss Davis explained how oppressive the partnership we had created in East St. Louis sometimes felt to her and her neighbors. She described how our students came to Emerson Park each week with the knowledge and skills that 15 credit hours of the best graduate school training in architecture, urban planning, public finance, organizational behavior, and environmental science could provide. Upon entering the community, they met her neighbors to engage in cooperative problem solving aimed at addressing the extremely complex environmental, economic, and social problems confronting the neighborhood. She reminded us that the overwhelming majority of the Emerson Park residents we worked with each week never had the opportunity to attend college and were disadvantaged when working with our students because we were not offering residents the kind of training needed to prepare them to engage in cooperative problem solving on a truly equal footing with our students. Richard Suttle explained, "Without real training in topics related to community planning and design, we are not even the tail on the dog in this partnership. We are not even the flea chasing the tail on the dog. We are the flealet chasing the flea hoping to land on the tail of the dog."

While acknowledging our enormous contribution, Miss Davis stated that we had inadvertently reproduced an unequal partnership that reflected society's unexamined racism, sexism, and classism. Stung by Miss Davis's words, my colleagues and I were speechless! She softened this "bad news" with the "good news" that we could address this problem by working with her colleagues to create a new adult education program in East St. Louis, one that would offer the city's expanding network of civic leaders an opportunity to acquire the same knowledge and skills that our best students possessed. (The free school that Miss Davis proposed reminded my colleagues and me of the Highlander Folk School established by Myles Horton in New Market, Tennessee, in the 1930s to promote political literacy among Appalachian and southern people.) Miss Davis closed by saying that she and her neighbors would not have raised these issues had they not felt that we shared a strong relationship. She also acknowledged how nervous some of her neighbors were about speaking truthfully to us as representatives of a powerful white institution because in the past such behavior had usually resulted in a withdrawal of resources. We thanked Miss Davis for her honest feedback and said that she and her colleagues had given us a great deal to think about.

My colleagues and I were in fact livid with Miss Davis and her neighbors when we left the neighborhood house. As we drove back to our campus, we exhausted most of the four-letter words we knew. Each of us had become involved in the East St. Louis Action Research Project above and beyond our regular university duties—some of us without the security of tenure. But as we approached the town of Effingham, an hour northeast of East St. Louis, we looked at each other and acknowledged with a laugh the fundamental truth of Miss Davis's criticism of our work.

Following this epiphany in Effingham, we held several meetings with the students and faculty to discuss the implications of the residents' critique and proposal. We decided to develop a "neighborhood college" that would annually offer three to four free courses on community planning and design topics. In the spring of 1996 I had the privilege of teaching the first course offered by the neighborhood college: principles and methods of direct action organizing. Forty-three neighborhood activists, social service providers, and municipal employees completed this course, which was designed to last 2 hours a week for 11 weeks. Resident interest in the course was so intense that our weekly sessions rarely ended in less than 3 hours, and as a result of student demand, the class was extended an additional 5 weeks. During the past 5 years, more than four hundred East St. Louis residents have participated in a total of twelve courses offered by the neighborhood college.

Accomplishments of the EPDC–UIUC Partnership and Emerson Park's Future

The year 1999 was a banner one for the university's 10-year-old partnership with EPDC.[1] Groundbreaking for the $30 million Parson's Place project took place. EPDC secured a highly competitive YouthBuild Program to train unemployed residents for jobs in the building trades. This program is being cosponsored by EPDC, the Lessie Bates Davis Neighborhood House, the local carpenters' union, and the UIUC. The Jackie Joyner-Kersee Foundation has opened a state-of-the-art youth recreation and development center within easy walking distance of the Parson's Place site. More recently, a local chemical plant that had been a major problem for the neighborhood was acquired by a Scandinavian company with an excellent environmental and labor record. In addition, this firm has decided to make its Emerson Park facility their North American headquarters and has already added scrubbers to their smokestacks and approached EPDC and Lessie Bates Davis house to discuss the development of a local job training program. EPDC's expanding budget has enabled the organization to increase its full-time staff from two to four persons. These and other accomplishments have earned EPDC and the university several important accolades.

The EPDC–UIUC partnership has produced plans that have led to more than $45 million in new public and private investment within this once-devastated neighborhood. It has also led to the transformation of EPDC from a volunteer association into a well-respected community development corporation. In addition, it has encouraged neighborhood leaders from the city's other residential neighborhoods to enter into similar collaborations with the university. This work established the groundwork for the creation of the university's Neighborhood Technical Assistance Center.

The partnership has also offered more than 3,500 UIUC students powerful learning experiences that have challenged their intellects, their hearts, and their souls. Life-transforming experiences working with Miss Davis (see figure 5.6), Mr. Suttle, and their Emerson Park neighbors have also prompted dozens of UIUC architecture, landscape architecture, and urban and regional planning students to pursue professional opportunities with community-based organizations serving low-income communities of color.

The EPDC–UIUC partnership serves as an optimistic antidote to the pessimistic social commentary of the 1990s. However, the most intense feeling I had during my years working in East St. Louis was rage at the unequal distribution of the fruits of one of our nation's longest periods of economic expansion. My rage is tempered by the heroic example of leaders such as Miss Davis, Mr. Suttle, Miss Haney, Miss Tucker, and Mr. Peete, whose faith has sustained them on a journey that began during the early days of the civil rights movement.

Despite all of the work that needs to be done in East St. Louis and other communities devastated by the increasingly powerful forces of our global economy, I am hopeful regarding the future. Every time a former student such as Richard Koenig quits a comfortable job with a large state housing agency to become the director of a struggling community development corporation, or Juan Salgado decides he can make his most important professional contribution working as a community organizer in a rapidly changing immigrant community, or Michelle

Figure 5.6
Miss Ceola Davis (center front), founder of the EPDC; Mr. Henry Peete (front right), EPDC secretary; and Mr. Paul Foppe, graduate research assistant, ESLARP. (Photograph by K. Reardon.)

Whetten, raised in suburban comfort, accepts a challenging community planning position in New York City, or Rafael Cestero returns to his hometown of Rochester, New York, to rebuild its neighborhoods from the grassroots up, I am encouraged. Their commitment to a planning practice that seeks to mobilize historically marginalized groups to address the structural factors causing increasing social inequality in our society makes me guardedly optimistic about America's future. This is the most hope one can muster in a nation where East Harlem children suffer rates of respiratory illnesses as high as those of children living in West Africa's poorest countries and where one of the leading causes of death for East St. Louis children below the age of five is opportunistic diseases related to AIDS.

Editors' Introduction to Chapter 6

Many (perhaps most) planners believe that the skilled application of professional knowledge and expertise can create sustainable places. Sandercock argues that creating sustainable cities requires instead a "therapeutic planning practice." While this claim that planners should practice a form of therapy may rattle some readers, there is ample evidence that many people fear strangers and that professional planners have learned to distrust, indeed fear, strong emotions. Over a decade ago Seymour Mandelbaum (1991) wrote of conventional planning practice as containing heated passions and redistributing them as a set of rational choices. Both fear and distrust lurk just below that contained public discourse, leaching out and transforming actions on the surface.

Hope is constitutive of Sandercock's concept of therapeutic practice, and as such it acts as an antidote to fear and distrust. Restating Antonio Gramsci's comment about the "pessimism of the intellect and optimism of the will," geographer David Harvey (2000, p. 17) calls for an "optimism of the intellect." Sandercock implicitly responds to that call by defining intellect inclusive of the emotions. For further insight into how embodied emotions influence professional practice, one could turn to Pierre Bourdieu's Pascalian Meditations *(2000) and Martha Nussbaum's* Upheavals of Thought *(2001), which are from the inquiry of cultural study and literature, respectively.*

My students chatted excitedly during the 3-hour trip; however, as our van approached the city, they grew very quiet. I watched them as they gazed at the city's blocks of abandoned industrial buildings, railroad yards, commercial areas, and single-family homes. By the time we arrived at the Lessie Bates Davis Neighborhood House, the van was dead silent.

—Kenneth Reardon

6

Dreaming the Sustainable City: Organizing Hope, Negotiating Fear, Mediating Memory

Leonie Sandercock

I look into my crystal globe, and I dream of the carnival of the multicultural city. I don't want a city where everything stays the same and everyone is afraid of change; I don't want a city where young African Americans have to sell drugs to make a living, or Thai women are imprisoned in sweat shops in the garment district where they work 16 hours a day 6 days a week; where boys carry guns to make them feel like men, suspicion oozes from plaster walls, and white neighborhoods call the police if they see a black or a stranger on their street. I don't want a city where the official in charge refuses to deal with the man standing at his desk because everything about him is different; where immigrants are called "blackheads" and forced to find shelter in the industrial zone; where whites pay more and more of their private incomes to protect themselves from "strangers" and vote for officials who will spend more of everyone's tax dollars on more law and order rather than more schools and health clinics; where political candidates run on promises of cutting off services to "illegal immigrants"; where the media teaches us to fear and hate one another and to value violence in the name of "patriotism" and "community." I don't want a city where the advertising men are in charge and there are no circuses for those without bread. I don't want a city where I am afraid to go out alone at night, or to visit certain neighborhoods even in broad daylight; where pedestrians are immediately suspect, and the homeless always harassed. I don't want a city where the elderly are irrelevant and "youth" is a problem to be solved by more control. I don't want a city where my profession—urban planning—contributes to all of the above, acting as spatial police, regulating bodies in space.

I dream of a city of bread *and* festivals, where those who don't have the bread aren't excluded from the carnival. I dream of a city in which action grows out of knowledge and understanding; where you haven't got it made until you can help others to get where you are or beyond; where social justice is more prized than a balanced budget; where I have a right to my surroundings, and so do all my fellow citizens; where we don't exist for the city but are wooed by it; where only after consultation with us could decisions be made about our neighborhoods; where scarcity does not build a barbed-wire fence around our carefully guarded inequalities; where no one flaunts their authority and no one is without authority; where I don't have to translate my "expertise" into jargon to impress officials and confuse citizens.

I want a city where the community values and rewards those who are different; where a community becomes more developed as it becomes more diverse; where "community" is caring and sharing responsibility for the physical and spiritual condition of the living space. I want a city where people can cartwheel across pedestrian crossings without being arrested for playfulness; where everyone can paint the sidewalks, and address passers-by without fear of being shot; where there are places of stimulus and places of meditation; where there is music in public squares, buskers (street entertainers) don't have to have a portfolio and a permit, and street vendors coexist with shopkeepers. I want a city where people take pleasure in shaping and caring for their environment and are encouraged to do so; where neighbors plant bok choy and taro and broad beans in community gardens. I want a city where my profession contributes to all of the above, where city planning is a war of liberation fought against dumb, featureless public space; against STARchitecture, speculators, and benchmarkers; against the multiple sources of oppression, domination, and violence; where citizens wrest from space new possibilities and immerse themselves in their cultures while respecting those of their neighbors, collectively forging new hybrid cultures and spaces. I want a city that is run differently than an accounting firm; where planners "plan" by negotiating desires and fears, mediating memories and hopes, facilitating change and transformation.

With the ideas in these two paragraphs (now expanded) I ended my book, *Towards Cosmopolis* (1998), which is in part an imagining of how we all, children of various diasporas, might manage to coexist in the shared spaces of our cities and neighborhoods across the planet. In this chapter, these sentiments are a starting point for thinking about the sustainable city in relation to the work of the city-building professions. We need to ask if there is such a future, such a city, such a profession. How do we get there, build it, and retrain ourselves?

I look around me, in my own city and on my travels, and what I sense is that people are searching for one or more of the following: diversity, community, and sustainability. These contemporary urban quests are the sources of the positive energy that can build sustainable cities in the third millennium. I begin by discussing each of these quests and how they might reinforce each other, then segue into what this means for planning and planners, concluding with a reflection on what all of this has to do with stories and storytelling.

The Search for Diversity

For the past 50 years, in British, North American, and Australian cities, we have been deliberately, through planning and design practices, fragmenting our cities into homogeneous, isolated, self-contained spaces—suburbs, shopping malls, industrial parks. The consequence is that most of us do not experience the complexities of the city directly and physically, in where we walk or whom and what we see.

My inner-city neighborhood in Melbourne, Australia, is unusual. In an otherwise predominantly suburban landscape, it is notorious for its diversity—a diversity that embraces a wide range of incomes, age groups, and ethnicities; housing tenure types; an indigenous population; street sex workers, wannabe and on-the-way-to-being rock stars, and assorted other alternative lifestyle folks; and a diversity of land uses, from residences to cafes, small-scale retail stores to music venues, beachfront parks, and municipal botanical gardens. Within 200 meters of my house there are four boarding houses. The analogy in the United States is a single-room occupancy hotel (SRO), but these local boarding houses are different. They are an unobtrusive part of the urban fabric in which dwell various mixtures of elderly, disabled, single employed younger males, single women with children, some unemployed folks, some welfare dependents, some indigenous people, and some very non-nine-to-five characters. Every day, at least three times a day, one such character, a man in his forties with long matted hair, and his equally unkempt dog, walk past my house. The man always carries a big stick, which he thwacks on the ground as he walks. I have seen him at close quarters in the liquor store, buying his bottles of beer while I bought my bottles of merlot. I look at him sideways. At home, in private, I call him "the Boogie Man." He features in my nightmares. And yet, it is clear to me that he belongs in this neighborhood, as much as any of us do. This man (and his dog) would not find accommodation in the suburbs. He would not be able to walk the streets of those neighborhoods without some anxious parent calling the cops. He is the strange angel who redeems the inner city and keeps alive the possibility of difference.

In 1998 the Melbourne metropolitan planning agency published a report on sociodemographic trends titled, "From Donut City to Cafe Society," which documented a reversal of locational preferences in the housing market that is beginning to have a significant impact on urban form. For the past decade Melbourne has been undergoing a transformation of its urban fabric and a reshaping of the pattern and direction of its growth as reurbanization occurs. After 40 years of relentless outward, low-density suburban expansion and the decline and hollowing out of the inner city (the donut city effect), there are now significant countertrends as young people (the 20- to 35-year-old demographic group) reject the suburban way of life in favor of the greater diversity of the inner city. Many of their parents are also selling the family home in the suburbs once their children grow up and are moving into inner-city apartments, town houses, or historic housing stock that was built at medium density. Parts of the inner city that had been previously rundown, neglected, or abandoned—along the river, the old bay shore neighborhoods, the docklands, even downtown itself—have suddenly been discovered as desirable places to live and have seen significant investment in renovation of older housing, conversion of industrial and office buildings to residential units, and the building of new, high-rise apartment buildings. A complicated mix of market forces, state policies, civil society's preferences, and social movements have driven this transformation.

The shift in housing type and location preferences has been accompanied by two other changes: a boom in retail, sport, entertainment, and cultural facilities in the inner city; and a vast improvement in the urban public realm as attention is paid to improving the quality of urban design of public space. While some urban analysts interpret this transformation disparagingly as "the city of spectacle" or the "city of consumption," and lament the growing inequalities between an increasingly affluent inner city and a neglected outer suburban realm, the reality is more complex. It is true that some of those who have moved into the inner city are primarily focused on a life of consumption. But it is also the case that the younger age groups want to live a different and less isolated life than that of their parents, and that this move back to the city embodies significant value shifts, including the desire for a more convivial, sociable life than is possible in the privacy-obsessed suburbs. This includes the desire to spend more time walking, cycling, and using public transport rather than sitting in cars, which for some reflects a concern about sustainability issues; the desire to be among a greater variety of people, in terms of age, ethnicity, and income, than is possible in the essentially homogeneous suburbs. It includes that less tangible but nevertheless real attraction of urbanity, the enjoyment of the city as *oeuvre* (Lefebvre, 1996), the playfulness, surprise, unpredictability, spontaneity—yes, the spectacle—of urban life, especially the uncommodified pleasures related to crowded streets, bright lights, energy, movement; the possibilities for anonymity as well as chance encounters; possibilities for learning just by looking and observing; the mixtures of stimulation and meditation.

This search for diversity and urbanity, expressed in the trends of reurbanization and revitalization of downtowns and older inner-city precincts, is not unique to Melbourne. It is happening in European and Canadian and some U.S. cities and it contains within itself some conflicting desires. Some of the new apartment buildings are high-security, vertical gated communities, whose occupants seem to be choosing to be in the city but not part of it, in terms of any sense of community. Others, however, become passionate defenders of the qualities of place that attracted them there, including the history and memory embodied in the urban fabric, as well as advocates of the social, cultural, and ethnic diversity of these neighborhoods. Some have pursued this goal to its necessary public policy conclusion, demanding some form of state intervention to ensure that the property market does not completely dictate who can enjoy the newly discovered urbanity. The opportunity and the challenge exists to link this search for urbanity even more clearly with the other positive energies in the contemporary city, and to think about how to build more diversity into existing and future suburbs.

The Search for Community

There is a sociological tradition going back to the middle of the nineteenth century that has both documented and lamented the loss of community with the onslaught of the industrial revolution. And yet, as powerful as those forces of erosion have been, there have been equally valiant struggles for the creation, recovery, or defense of community, worldwide. Even in the face of globalization, the life space of the local community has continued to assert itself and its claims. In those cities that attract transnational or rural migrants, the membership of any one geographically defined community may change dramatically between one census period and the next, posing the challenge of creating and building community as well as defending it.

"Community," though, is a slippery concept and a complicated reality, and although it has long been the apparently unassailable icon of much radical and utopian thinking, it needs to be critically examined. One starting point is the observation that "community" actually means different things in different contexts to different people. In contemporary America there is a tendency to think of it as bottom-up, anticorporate, in defense of life space; the sharing of a specific heritage or common culture and set of norms; embodying face-to-face relations and a more participatory politics. However, community can equally mean "my region or culture against yours"; my white Australian or Canadian or French culture against your foreign culture; my German or French culture against your Turkish or Algerian culture; or my true or authentic or local working-class neighborhood against your inauthentic or cosmopolitan or aestheticized middle-class one. There is a long tradition of community-based movements that have been primarily inspired by exclusionary sentiments, whether based on race, ethnicity, religion, or class. Richard Sennett (1970) has discussed how a "myth of community" operates perpetually in American society to produce and implicitly legitimize racist and classist behavior and policy.

However, there is more than one ideal of community, and Cornell West and bell hooks offer an alternative to the exclusionary model in the following passage:

It is important to note the degree to which Black people in particular, and progressive people in general, are alienated and estranged from communities that would sustain and support us. . . . We confront regularly the question: "Where can I find a sense of home?" That sense of home can only be found in our construction of those communities of resistance bell [hooks] talks about and the solidarity we can experience within them. . . . As we go forward as Black progressives we must remember that community is not about homogeneity. Homogeneity is dogmatic imposition, pushing your way of life, your way of doing things onto somebody else. That is not what we mean by community. . . . That sense of home that we are talking about and searching for is a place where we can find compassion, recognition of difference, of the importance of diversity. (hooks and West, 1991, p. 18).

Carlo Rotella (chapter 4 in this volume), revisiting his once-diverse old neighborhood of South Shore, Chicago, spoke of the folks who bought his parents' house when they moved out, describing them as "the last white couple to buy a house" in that part of the South Side. That couple, as Rotella tells it, were motivated by a dream of intercultural coexistence, of community building, yet they were ultimately driven out by fear. Now, a new dream of South Shore living, elaborated in the developer's brochure for LakeShore Pointe, talks of a friendly and secure neighborhood. The developer's rhetoric seeks to write a new story for South Shore in order to seduce buyers back into the area. However, the developer's story is not really one of community building (or rebuilding) in the sense that I am advocating.

Community, in the sustainable city, has to be inspired by the values expressed in hooks and West's writings and by Mel King's (1981) work in Boston and beyond. For King, the whole purpose of our struggle, past, present, and future, is to "create community," by which he means the human context in which people can live and feel nurtured, sustained, involved, and stimulated. "Community is the continual process of getting to know people, caring and sharing responsibility for the physical and spiritual condition of the living space" (1981, p. 233). His work is informed by a vision of the city we can build together, but he also knows that we are not there yet, and that moving toward this vision "will entail some very uncomfortable experience as we root out our prejudices and confront our fears" (1981, p. 240). I am particularly interested in King's belief that "we need, as individuals and as communities, to be about getting people to deal with the fears which immobilize us and bar us from our basic instincts towards growth, change, and harmony" (1981,

p. 232). I will return to this idea that the search for community requires us to deal with fear, since the very attraction of community is its potential sense of belonging; however, the dark side of that desire for belonging is its potential to exclude others who are deemed not to belong. The search for community is very much alive in the cities that we are all familiar with, but implicit within this search are negative as well as positive energies. The challenge is to know how to bring out and to work with the positive energies.

The Search for Sustainability

Part of the positive energy of creating community is the notion of caring about others and sharing responsibility for the physical and spiritual condition of the living space. The search for ecological sustainability is becoming, and must become even more, a city-focused quest. This case has been persuasively made by, among others, Peter Newman and Jeff Kenworthy (1999), who argue that a sustainable city must be a city of urban villages. It must be a compact city of far greater density than is the norm at present in Australia and North America, and it must be a city that has overcome its dependence on the automobile. The first step in their vision of the sustainable city is the revitalization of the inner city, which has already begun and has usually been associated with community processes that have developed a new vision for an area. (This is the metanarrative of Reardon's chapter, to which I will return.) This vision is sometimes associated with strategies of housing rehabilitation, sometimes with preservation of historic buildings and streetscapes, with street festivals and other community arts events. Sometimes it is associated with the provision of low-income housing to retain a mix of incomes, and investment in new businesses by innovative entrepreneurs. Finding the right spark for regeneration requires creativity and commitment by planners and urban managers, but it also always requires significant community input and effort and is often triggered by a community's mobilization in defense of its life space. Reardon's students' first response to Emerson Park in East St. Louis was to want to run away, but they were encouraged by the faith and commitment of the residents. Faith is an important word here. So is hope.

Newman and Kenworthy (1999) note that while there are now strong market forces pushing inner-city revitalization in some cities (such as the development of LakeShore Pointe in Rotella's story), revitalization alone will not attract people back into the city if the streets are not safe and the schools are not satisfactory. One of the major themes in Reardon's story is the effort to make the streets safe again for children to walk to school. In East St. Louis, Reardon's project for 10 years, this revitalization was initially achieved by the literal sweat of residents, along with student helpers. Once results were visible, the public funding agencies began to take an interest. So faith, hope, and sweat were the catalysts. Once begun, this improvement in the public environment became a signal and a catalyst for broader regeneration.

Newman and Kenworthy's sustainable urban future proceeds from inner-city regeneration to taking those inner-city qualities, the diversity and urbanity I described earlier, to the suburbs and developing "urban villages" in the suburbs; that is, dense concentrations around public transport nodes. Bernick and Cervero (1996) have provided examples of such transit villages from around the world, including a growing list in the United States, and Calthorpe (1993; Calthorpe Associates, 1990) has drawn up design guidelines for public transit-oriented design, as have the New Urbanists (Katz, 1994). This vision of sustainable cities composed of urban villages is not just compatible with but in fact depends on both the search for diversity and the search for community in that it envisages neighborhoods that are not only mixed, vibrant, and playful, but which are also communities that take more control of their living space and promote the broadest possible sense of belonging. There are many inspiring examples of such efforts, from the well-known Brazilian city of Curitiba to European cities like Zurich, Stockholm, and Copenhagen, to Boulder, Colorado, and Portland, Oregon, in the United States and Toronto and Vancouver in Canada. These are examples that the authors describe as "positive city-building processes rather than the city-destroying processes of dispersal, pollution, and community disturbance associated with automobile dependency" (Newman and Kenworthy, 1999, p. 181). Newman and Kenworthy quote Jan Gehl, who describes how over a 20-year period Copenhagen began to win back its city:

By the 60's American values had begun to catch on—separate isolated homes and everyone driving. The city was suffering, so how could we reverse these patterns? We decided to make the public realm so attractive it would drag people back into the streets, whilst making it simultaneously difficult to get there by car. (Newman and Kenworthy 1999, p. 204)

Each year Copenhagen reduced central area parking by 3 percent; converted more streets to pedestrian traffic only; built or refurbished city housing; and sponsored all kinds of improvements of the public space, from street furniture and sculptures to more buskers, markets, and other street life. As Gehl said, "the city became like a good party" (1992, p. 204), and as a result, there is now a declining market for single detached homes on the urban fringe. When the city becomes like a good party, we have achieved the ludic city about which Henri Lefebvre wrote so eloquently (Lefebvre, 1996).

So far, I have outlined three components of the sustainable city. I hope it is clear that this is not urban science fiction or wishful thinking in that the seeds of these changes have all been planted and there are enough demonstration projects in a wide range of cities and neighborhoods to provide inspiration. The role of community groups and community mobilization has been critical in redefining a vision of the good city. The market is responding where it sees a profit to be made, most notably in inner-city housing and consumption-based activities. What of the role of planners and planning? What needs to change in professional practice?

Toward a Therapeutic Planning Practice

I have written elsewhere (Sandercock, 1998) of the shift from a modernist to a postmodern paradigm for the city-building professions, and Newman and Kenworthy (1999) also emphasize this in their discussion of how to move toward the sustainable city. They talk about the need to develop new sets of technical standards, manuals, and regulations, as the New Urbanists have begun to do. Their appendices provide a range of useful tools, from growth management approaches and guidelines to a checklist for city sustainability using economic efficiency, social equity, environmental responsibility, and human livability criteria, to a draft economic impact statement for urban development. Finally, they provide a set of strategies for improving public transportation and land use integration in automobile-oriented cities. These are no doubt essential, but I want to focus now on something that is possibly even more important. This relates to the language we use as planners and how that language constrains the kind of work we might do to bring about transformations of values and institutions.

The language traditionally used in planning practice has been a rational discourse that explicitly avoids the realm of emotions, which is of course the stuff of storytelling. If we think about that for half a minute, it is an extraordinary and bizarre feat to talk about diversity or urbanity without talking about memory, desire, spirit, playfulness, eroticism, and fantasy. Or to talk about community without talking about longings and belongings, losses and fears, guilt and trauma, anger and betrayal. Or to talk about sustainability without talking about hostility and hope, compassion and caring, greed and nurturing. There is an ethics of city life and city death, a series of both everyday and long-term choices that get made and reproduced, and there is a corresponding ethical language in which to discuss such choices. There are desires and fears bound up in community building and a corresponding language of emotional acknowledgement. I want to suggest that more and more of our efforts, if we want to work toward sustainable cities, will be bound up with organizing hope, negotiating fears, mediating memories, and facilitating community soul searching and transformation. This means we should take to heart what Mel King has been practicing for 40 years in Boston, which is getting people to deal with the fears that immobilize them and the prejudices that are both part of yet also distort our basic humanity. I would like to talk about these activities (that is, about a more "therapeutic practice") in the remainder of this chapter.

The Organization of Hope

Anastasia Loukaitou-Sideris (2000) and her students at the University of California-Los Angles recently concluded a participatory action research project in the Pico Union inner area of Los Angles. This area is home now to large numbers of Mexican and Central American immigrants and refugees and has experienced massive decline and disinvestment. Part of the approach used was to research and record the multicultural social history of the area and then request a name change, from Pico Union to the Byzantine-Latino Quarter, as part of the process of creating a sense of identity, past, present, and future, to assist with community building. This in itself is an interesting approach. The name change begins telling a new story about the neighborhood, one that is redolent of pride in a diverse past and hope for a sustainable future, which depends on the exercise of community building. Their next step was to begin to work on the rehabilitation of small neglected public spaces that scarred the area, using the community's labor and involving children. Then they sought permission from and support of various public authorities whose own neglected spaces were contributing to the overall effect of neglect and abandonment. This is a story in progress, but it is also an example of turning around despair and passivity through community action.

Ken Reardon's work in East St. Louis is probably the most successful and inspiring model we have in this country of what I call "the organiza-tion of hope." [This compelling phrase comes from the title of Howell Baum's (1997) fine book on community planning.] In chapter 6 in this volume Reardon tells the story of a 10-year university and community partnership that has a metanarrative and a number of micronarratives. Reardon's final reflection begins with rage, his overwhelming and ongoing feeling about living conditions in East St. Louis. However, the metanarrative is the following, "My rage is tempered by the heroic example of leaders such as Miss Davis, Mr. Suttle, Miss Haney, Miss Tucker, and Mr. Peattie, whose faith has sustained them on a journey that began during the early days of the civil rights movement."

The key words are "heroic example" and "faith." Reardon has given us an old-fashioned story for a postmodern age; a story of heroism against impossible odds, with faith and quiet determination as the weapons, an inspirational story. What it tells us about the work of plan-ners is that what they do is to try to organize hope. The micronarratives are the details, the insights, the lessons learned along the way, the most powerful of which is the "epiphany at Effingham," because it involves humiliation, an unmasking of the allies in the struggle as also oppressive partners. This moment of self-knowledge is another classic ingredient in heroic stories. Other lessons include the importance of participatory action research methods; of removing the fear of violence, which acts as a barrier to broader community participation; of securing the involve-ment of state and federal agencies; of ways of using the media strategi-cally; and finally, of luring private-sector investment back into the neighborhood. Each of these micronarratives is driven by the metanarra-tive, the overarching lesson, which is that none of this would have hap-pened without the faith, hope, and sweat of the quietly determined leaders and residents of East St. Louis.

The result, a decade later, is that more than $45 million in new public and private investment has come to this once-devastated neighborhood and in the process more than 3,500 University of Illinois (Urbana-Champaign) students have had a powerful, and for some, life-transforming, learning experience. The facts of Reardon's story range from the initial descriptions of devastation to the final triumph of securing $30 million of private investment. The plot is the step-by-step approach to regeneration. The tools are flow charts, Excel files, wall maps, interview schedules, and sweat equity. However, the deeper meaning of the story is its inspirational quality, its description of a process of organizing hope.

Negotiating Fear

Harvard Law Professor Gerald Frug (1999) argues in his new book, *City Making*, that the single most important issue facing America's cities is fear, fear of strangers; and, yes, specifically, fear of the black stranger, racialy based fear. I have also written about how the future of planning in polyethnic or multicultural societies requires a coming to terms with the existence of fear in the city, fear of the other, the stranger, foreigner, outsider (Sandercock, 2000a,b). Existing planning approaches dealing explicitly with fear in the city revolve around crimes to property and threats to personal safety, and have focused on either urban design (Oscar Newman's "defensible space") or environmental design (the crime prevention through environmental design literature). That is, they deal with the hardware of crime prevention rather than the software of fear itself.

The recent emphasis on more "communicative approaches" for handling planning disputes (Innes, 1995; Healey, 1997) acknowledges the need for more process-based methods of conflict resolution, but their emphasis on rational discourse avoids the emotions at the heart of conflict and thus often avoids the real issues at stake. I want to suggest a more therapeutic approach, which begins with an analysis and understanding of this fear of the other and develops solutions that are really processes for confronting these fears.

To illustrate this potential I give here a brief account of a recent cross-cultural conflict over land use in inner Sydney and its resolution through a therapeutic process in which a space was created for speaking the unspeakable, for talking about fear and loathing as well as hope and transformation. The issue concerned the future of a factory site immediately adjacent to the residential area known as the Block in the inner Sydney neighborhood of Redfern. The Block had been a 1970s federal government initiative that had granted a parcel of inner urban land to aboriginals. This area has received a lot of media and political attention in recent years as housing owned by the Aboriginal Housing Corporation deteriorated and the area became a center of drug dealing and use. Local opinion was divided about the aboriginal presence, with some nonindigenous locals believing, or hoping, that the government would "clean up" the area before the Olympics, while other nonindigenous residents were firmly committed to a multiracial neighborhood as a symbol of a wider reconciliation process in the nation.

In the 1980s the local council had rezoned the site for community use, which meant that when the factory closed down a decade later, the council had to acquire the site. It then tried to rush through an approval to demolish the buildings on the site, in sympathy with the conservative white residents' faction, who wanted the site to become a park with a police station at its center. This group expressed strong disapproval of any use of the site for aboriginal purposes. The indigenous community wanted the buildings and site used for aboriginal economic and community purposes, including a training facility. A third group, white residents holding broader reconciliation values, supported the aboriginal group and the larger issue of a continuing aboriginal presence in the area.

After being embarrassed by residents' protests, the council backtracked and hired a social planning consultant, Wendy Sarkissian,[1] to conduct a consultation process that would result in recommendations for a master plan for the 2,200-square-meter site. The consultant's initial scoping of the situation suggested to her that there was such hostility between the three identifiable groups of residents that any attempt at a general meeting to start the process would either meet with a boycott from one or more groups or end up in violence. Her strategy therefore was to organize a series of discussions, to listen to people's hopes and fears. For the first few months, separate meetings were held with each of the three "camps." These included small meetings in people's living rooms, larger meetings in more public settings, meetings with children, and meetings with members of the aboriginal community using a black architect as mediator.

After all this preparation, which, on the part of the planning team mostly consisted of listening, a "speak-out" was organized in which each group agreed to participate. This was by far the most risky part of the whole process in that it was the most likely to get out of hand. The point of holding such an event is to allow people to say what they feel, no matter how unpleasant or how painful it might be for others to hear. The hope implicit in such an event is that as well as allowing the unspeakable to be spoken—that is, performing a sort of cathartic function for all those carrying anger or fear or betrayal inside themselves—the words would also be heard, in their full emotion, by those whose ears and hearts had previously been closed. The consultant's intention was to encourage the real issues at stake to be aired before any site-specific discussions.

The real issues ranged from resentment on the part of conservative whites at the aboriginal presence in "their" neighborhood, to concerns about personal safety and children's well-being related to the presence of drug dealers and users, to, on the part of indigenous people, anger and sadness at 200 years of domination by "white fellas" who even now had little understanding of their history and culture. At one point in the speak-out, the consultant herself was verbally attacked by a tearful aboriginal woman storyteller who demanded to know how the consultant thought she could change 200 years of racist history in a few months, with a few meetings. There is no satisfactory answer to such a profound question, only the honest answer in this case, which was an attempt to create a space, in one place, at one point in time, where perceptions might shift, where public learning might occur and some larger transformation take place.

Such a shift did in fact take place. It is interesting that before the speak-out, the consultant had been criticized by white residents sympathetic to indigenous desires for her overly therapeutic approach, for too much talk about feelings. Clearly what had been happening during the initial period of meetings and listening was the creation of a safe space in which parties could meet and speak without fear of being dismissed, attacked, or humiliated. The speak-out would not have been possible without this preparatory work, which also involved the building of trust in the consultant and her team. The speak-out itself also had to be designed as a safe space, and this was achieved in part by ceremonializing the activities.

In describing the mediation practice of Shirley Solomon, working on a dispute between Native Americans and a county government in the state of Washington, John Forester notes that "the ceremonial design of innovative public policy conversation can be an important signal to all parties that they are about to engage in a different—fresh and non-threatening—kind of exploratory conversation in a different, deliberately designed setting" (Forester, 2000, p. 151). Solomon moved from the first stage of creating a safe space to a second stage, creating a sacred space in which the whole idea was of "getting to higher ground" (Solomon, in Forester, 2000, p. 152). Solomon's story, like Sarkissian's, teaches us not only about caution in the face of explosive histories but also about the place of storytelling in setting the stage for the beginnings of reconciliation. Neither of these practitioners wishes to disguise the fact that "all this is experimentation. It's not like there's a cookbook, and you're following it, and it all goes the way it should" (Solomon, in Forester, 2000, p. 158).

After the speak-out, it was possible to move the process on, to hold joint group discussions and negotiations, to forge a set of principles for deciding the future use(s) of the site, and finally to hold a set of meetings to draw up guidelines to present to the council. One of the principles guiding the consultant was the determination not to force closure before there was the possibility of a genuine agreement, rather than a mere "deal," an unsatisfactory compromise. That agreement finally came, after 8 months and an expenditure of about $50,000. The outcome, ten guidelines for a master plan, was a compromise, but it was also a breakthrough of sorts, in that the white conservatives backed off from their opposition to *any* aboriginal use of the site and agreed to some training facilities.

This kind of planning work, involving confrontation and dialogue and negotiation across the gulf of cultural differences, requires its practitioners to be fluent in a range of ways of acquiring information and communicating: from storytelling to listening to interpreting visual and body language. It would seem to be a model that is relevant to the new complexities of nation building and community development in multicultural societies. It is probably, the best model in situations where direct, face-to-face meetings are unthinkable or unmanageable because of histories of conflict and/or marginalization. In such cases, in carefully designed public deliberative processes, the use of narrative, of people telling their own stories about how they perceive the situation, becomes a potential consensus-building tool for uncovering issues that are unapproachable in a solely rational manner. When the parties involved in a dispute have been at odds for generations, or come from disparate cultural traditions, or where there is a history of marginalization, something more than the usual toolkit of negotiation and mediation is needed, some method that complements but also transcends the highly rational processes typical of the communicative action model. In the case just discussed, that "something more" was the speak-out, which provided an occasion for dealing with history in highly personal, narrative, and emotional ways.

There are other possible methods—using drama, for example—or other more symbolic or nonverbal means of storytelling and communicating deeply felt emotions. Indigenous people and other long-dominated groups are often, with good reason, preoccupied with the unacknowledged and therefore unfinished business of the past. It is particularly important for them to be able to tell their stories. However, all parties involved in planning disputes have a story, and there is growing recognition of the importance of the telling and hearing of stories in conflict resolution. Narratives about the past can be vital in navigating long-standing, cross-cultural disputes.

Norman Dale (1999), a Canadian planner experienced in working through conflicts involving First Nations, argues that stories are a common denominator valued by all ethnicities and ages. "Relatively few people learn the rules of specialized modes of discourse such as legal argumentation or Western scientific debate. But nearly all of us, beginning as very young children, are immersed in stories—whether of fiction or family tales" (1999, p. 944). A more democratic and culturally inclusive planning not only draws on many different ways of knowing and acting, but also has to develop a sensibility able to discern which ways are most useful in what circumstances. What is currently missing in most of the collaborative planning and communicative action literature is this recognition of the need for a language and a process of emotional involvement, of embodiment, of allowing the whole person to be present in negotiations and deliberations.

There are, however, two notable exceptions to this—Howell Baum (1997, 1999) and John Forester (1999)—and I turn to their work in the next section. It is sufficient to note here that Baum argues that when such language and behavior is disallowed or discouraged by planners insisting that participants be rational or that discussions follow a logical order, they will elicit only superficial participation. "Told to be rational, people assume they have been told not to be themselves. They may feel relieved. Planning will not require them to reveal or risk what matters" (Baum, 1999, p. 12). Baum also argues that it is important for planners working in emotionally charged situations not to try to suppress conflict (a natural enough impulse, since most of us find conflict uncomfortable), for to do so is to sabotage the work of grieving and healing that needs to be done as part of a process of change; and that helping people to discuss their fears is a way of seeing past these fears toward the future.

What also interests me about the philosophy underlying this therapeutic approach is the possibility of transformation, of something beyond a merely workable tradeoff or Band-Aid solution. Much of the negotiation and mediation literature, argues Forester (1999), remains economistic, more concerned with trading and exchange than with learning, more concerned with interest-based bargaining and "getting to yes" than with the broader public welfare. However, just as in successful therapy there is a breakthrough and individual growth becomes possible, so too, with a successful therapeutically oriented approach to managing our coexistence in the shared spaces of cities and regions, there is the capacity for collective growth. Or, to move from the language of therapy to that of politics, there is the possibility of social transformation, of a process of public learning that results in permanent shifts in values and institutions.

My argument is that this approach is the only model that works in cases where histories of conflict have made more traditional negotiation techniques irrelevant. The conflict in inner Sydney was ostensibly over the fate of a 2,200-square-meter site in one municipality. The emotional and symbolic stakes, however, were considerably greater, especially for indigenous people, whose continuing presence in the inner city seemed to be the hidden agenda and longer-term threat. So this small, localized conflict was, in the minds of most of those involved, about core issues of history, injustice, and national identity. At whatever scale such conflicts arise—neighborhood, city, or region—there is a strong case for using this therapeutic approach. Indeed, South Africa's experience with its Truth and Reconciliation Commission points to possibilities of its application at the level of a whole society, in extraordinary circumstances.

Mediating Memories

Another dimension of the work of building sustainable cities and communities involves a process of mediating memories. All neighborhoods have histories, and that accumulation of history is constitutive of local identity. Part of the work of community building involves invoking this history, these memories. What many who do this work do not talk about, however, are those situations in which it is assumed that there is only one collective memory of place when it is more likely that there are layers of history, some of which have been rendered invisible by whomever is the culturally dominant group. There is now some fine work being done by preservationist planners such as Gail Dubrow (1998) and Dolores Hayden (1996) in exploring and acknowledging conflicting memories of place and mediating those memories in community planning processes in order to provide a more inclusive foundation for creating visions of a future community.

Howell Baum (1997) discusses a case in Baltimore where planners working with the Jewish Community Federation tried to suppress the conflict among members of the federation about the authority of different memories and the legitimacy of alternative futures, in order to plan harmoniously. Instead, the conflicts, undiscussed and unaddressed, persisted and the organization could not plan realistically. Likewise, John Forester's recent work shows the importance of dealing with the past; of explicitly addressing traumas and grievances, feelings of loss, anger, betrayal, of "not leaving your pain at the door" (Forester, 1999, p. 201). Baum emphasizes that the planning process must create a transitional space between past and future, "where participants can share the illusion of being apart from time. They need to imagine stepping away from past memories without feeling they have lost their identity or betrayed the objects of memory. . . . They must be able to imagine alternative futures without feeling obligated to enact any of them" (1999, p. 11). This is what he calls the "serious play" of a good deliberative planning process.

An interesting example of recognition of the need to deal with memory in order for reconciliation and healing to occur comes from Liverpool, England, a city that by the 1980s, after two decades of economic decline, was on the brink of city death, with astronomical levels of unemployment, corresponding outmigration of young people, appalling race relations, and a hideously deteriorated and neglected built environment. How can a city recover from such despair and demoralization? There were apparently two catalysts. According to Newman and Kenworthy (1999), the first was a community mobilization around housing rehabilitation. The second was a major effort to combat racism, "including removing this cancer from the police force . . . , providing special opportunities for those from the black community, . . . starting an arts antiracist program; and perhaps of greatest spiritual and symbolic impact, opening the Museum of Slavery in the new Albert Dock tourism complex" (1999, p. 328). This award-winning museum shows how Liverpool was central to the slave trade. It graphically depicts the whole process of slavery, and names the many established Liverpool families who made their fortunes from slavery. "The message for other cities caught up in inner city decline based on problems of race and dispirited neighborhoods is encouraging. The implications for sustainability are obvious" (Newman and Kenworthy, 1999, pp. 328–329). Here then is a case where the telling of a buried story or stories provides some ground for healing a divided city or neighborhood, and in so doing acts as a catalyst for regeneration and growth.

Conclusion: Telling Stories

There are a thousand and one urban reconciliation and regeneration stories and urban ecological dreams that can be dreamed, and a thousand tiny empowerments that can emerge by pursuing each of these dreams. This is how we build the sustainable city. It is grounded in and inspired by some exemplary models and practices, some of which are documented in this volume. It acknowledges that community mobilizations are generally the catalysts for change, but it also recognizes that more coordinated action among state, market, and civil society is necessary to move toward a sustainable city. Above all, it acknowledges that the work of planners involves both the creation and the use of stories and storytelling.

During the June 2000 symposium on storytelling, Ken Reardon explained how he addressed an audience of developers and real estate folks. He described the careful crafting of a speech. "I designed my presentation to highlight the dramatic steps that had been taken by neighborhood residents and local, county, state, and federal officials to improve East St. Louis's investment climate." He then recounted for the developers some recent success stories, followed by a description of investment opportunities. He finished with an emotional appeal to the audience, the need for a "few good men and women" to come forward and show some of the same faith that the residents and then state agencies had shown, by putting their capital on the line. This is a good example of what Throgmorton (1996) calls "planning as persuasive storytelling." Reardon told the audience of financiers a story of how the East St. Louis community had transformed itself through self-help, a morality tale bound to appeal to such an audience. Then, emphasizing how local and state agencies had already responded to this self-help effort, he in effect challenged the financiers to act with equal decency by showing some faith in the area as an investment opportunity. In a clever rhetorical strategy, morality and sound economics were now cozy bedfellows. Reardon both uses stories and creates a new story for East St. Louis, one defined by hope and faith.

In Liverpool a buried and shameful story was told as part of an attempt to address racism, as a precondition for moving the community out of its downward spiral. Anastasia Loukaitou-Sideris and her students' work in Pico Union shows how naming can create a new story that becomes the basis for neighborhood regeneration. (Equally, in settler societies, renaming the landscape begins to acknowledge indigenous realities.) Sarkissian's work in inner Sydney uses the telling of stories as a consensus-building technique to overcome fear and hostility. Reardon's metanarrative is an inspirational story, a spur to action. Barthel's work (chapter 11 in this volume) uses individual life stories in court to build up a picture of disadvantaged lives shaped by socioeconomic and racial inequalities. In his case, such "storytelling" can be literally lifesaving.

There are in fact multiple uses of stories and storytelling in the world of planners and planning. We have seen glimpses of storytelling as a research method and as a community participation technique, as in "tell me about . . .". We are enriched by Forester's work, which uses the stories of practitioners to create a deeper understanding of what it is that planners really do. We ought to be using storytelling techniques more when we write up (as government or consulting reports) the deliberations that have gone into community consultations or when we describe or craft the resulting community vision. I am always amazed at how dry such reports usually are, with their bulleted lists and numbered paragraphs, in spite of the fact that they are dealing and have dealt with the most passionate of emotions: loss, betrayal, anger, joy, and so on. Postcolonial critics and theorists like Linda Tuhiwai Smith, a Maori woman, advocate stories that celebrate survival (Smith, 1999). In a similar vein, in Australia Stuart Rintoul has gathered oral histories of black Australians in *The Wailing: a National Black Oral History*, which passes on beliefs and values and memories of injustice, rage, dignity, and sorrow.

We move back and forth between critical and narrative modes. We use stories as tools to keep memory alive, to celebrate our history or identity, to elicit lessons about how to act effectively, to inspire action, and as tools of persuasion in policy debates. As a general rule, the more entertaining the story, the more effective it is likely to be.

How, then, can we not conclude that stories and storytelling—in their various incarnations, as film, as mural, as rap or folk music, as dance, as witness, and yes, even as research reports—have a profound place in the work of the city-building professions as we try to build more sustainable cities for the twenty-first century. Stories are telling.

Editors' Introduction to Chapter 7

As a freshly trained planner, Berkshire began working on one of the most difficult challenges a professional planner can face: trying to find a site for a new landfill. One can read his tale as a familiar not-in-my-backyard (NIMBY) story and interpret it in the context of the extensive literature about conflict resolution, such as Lawrence Susskind and Patrick Field's Dealing with an Angry Public *(1996) and Susskind et al.'s* The Consensus Building Handbook *(1999). However, his city's backyard is not the only geographic scale, the only chronotope, that appears in his narrative. Other chronotopes from other geographic scales intersect the temporal movement of his tale and complicate the premises that customarily define the boundaries of "backyard." The intrusion of multiple geographic scales into the narrative challenges us to rethink the familiar traits that define the genre of NIMBY story. Eckstein's discussion of the spatial turn in narrative scholarship, Throgmorton's explication of tenticular radiations and remote effects that redefine the boundaries of one's home place, and Soja's insistence on the geographic imagination as an equal partner with the historical imagination all speak to the surprising ways in which Berkshire's seemingly predictable story unfolds in time.*

If it is read as altered by synchronous geographic intersections, Berkshire's story opens the door to new possibilities for waste management. For example, as Wolfgang Sachs et al. report in Greening the North *(1998), Germany and other European countries have made considerable progress toward reducing waste generation by turning to "ecological modernization" and "industrial ecology" and by requiring producers to assume responsibility for the disposal of their own products.*

If we also try a resistant reading of Berkshire's "practice story" (Forester, 1999), we can question his insistence that it was "our" job to "protect the board" and other decisionmakers. His own difficulty in expressing his "practice story" demonstrates the limits of pursuing sustainability under those professional terms and within the context of territorially bounded representative democracy. While Beauregard urges an alternative emphasis on discursive democracy, the difficulty Berkshire expresses raises questions about the viability of the version of discursive democracy attempted in his city region. Would Susskind et al.'s rational approach to resolving conflict or Sandercock's therapeutic practice have better enabled Berkshire and his Bluestem colleagues to negotiate the fears and distrust they encountered?

I want a city that is run differently than an accounting firm; where planners "plan" by negotiating desires and fears, mediating memories and hopes, facilitating change and transformation.

—Leonie Sandercock

7

In Search of a New Landfill Site

Michael Berkshire

Becoming a Regional Solid Waste Planner

With a bachelor's degree in business administration, I found my first job, in the mid-1980s, in a discount brokerage office located in a bank in Mount Pleasant, Iowa.[1] But like a lot of people tired of the policies of the Reagan administration, I spent more of my 3½ years at the bank thinking about environmental issues than trying to make more money for people. So I decided to find a career where I could apply my interests. I discovered urban planning.

As a planning student I had to complete a field problems group project for a client. My client was the East Central Iowa Council of Governments (ECICOG) and its regional solid waste planning coordinator, Liz Christiansen. After graduating, I was hired by ECICOG as their solid waste educator. I spoke to schools and civic groups, and I designed and implemented educational programs about waste reduction and recycling. I later became the regional solid waste planning coordinator for ECICOG, updating the regional solid waste comprehensive plan, which describes the waste reduction and recycling activities, programs, and services in ECICOG's six-county area.[2] I was applying my environmental interests.

One of the largest solid waste management organizations in ECICOG's region is Bluestem Solid Waste Agency. Linn County and the City of Cedar Rapids created it in 1994 by exercising a "28E" (joint powers) agreement. That agreement authorized Bluestem to provide environmentally responsible and cost-effective solid waste management for Linn County, in which Cedar Rapids, one of the largest cities in Iowa, is located. Bluestem's governing board consists of nine officials, six of them appointed by members of the Cedar Rapids City Council, and the other three appointed by the county board of supervisors.[3] Shortly after it was formed, Bluestem realized that the county's landfills were reaching capacity and decided to hire a planner to coordinate a process in response to this need. They hired me as this solid waste planner. It looked as though I could continue to apply my environmental interests to my job.

Searching for a New Landfill Site: The Process and the Public

Shortly after I was hired, Bluestem staff recommended to its board of directors a process in which alternative technologies available for managing solid waste would be investigated before deciding whether a landfill would be an appropriate component of an integrated system for managing solid waste in Linn County. We began by assembling a focus group of twenty-three Linn County citizens comprised of people with a lot of different experiences, interests, and knowledge. We met with that amiable group for almost a year. Given Bluestem's mission, we talked a lot about "integrated solid waste management," an industry term that means you try to appropriately match technologies with your actual waste stream. You look at how much is recyclable, how much is compostable, and you try to match the appropriate technology with that. The staff expended a lot of energy in the focus group meetings educating the group on what type of waste is generated in Linn County: how much of it comes from business and industry, how much of it comes from residents, and within those waste streams how much is plastic, and so on. You have to know what is actually being generated because it usually is specific to an area. For example, Linn County is a highly industrialized county that generates a lot of organic waste because of its corn-processing facilities.

After the focus group had an understanding of the waste stream, we started talking about the different technologies. We looked at every appoach from waste-to-energy, to transfer hauling, to putting the waste on a train and sending it to Utah. We then asked the group to recommend to Bluestem's board an integrated system for managing solid waste. In order to answer that question, the focus group went through an extensive "homework assignment" in which they looked at the particular waste stream, all the different technologies, and applied a complicated matrix in which one component would not conflict or compete with another component. (For instance, a municipal solid waste composting facility probably would compete economically with a typical yard waste composting facility.) They ended up recommending to our board a system that included continuation and expansion of the county's recycling and composting efforts, and building a facility that would manage household hazardous waste. The last component was a new landfill.

There were a variety of good reasons for choosing a landfill as the component that would take care of everything that could not be composted, recycled, or reduced at the source. One reason was that the group felt that the wastes generated in Linn County should be managed in Linn County; that Linn County citizens should take responsibility for the waste that is generated locally. They were also concerned about the loss of control if the waste would be sent to Illinois or outside the county. And they were concerned about the costs associated with other technologies. Waste-to-energy is expensive and so is transfer hauling. The focus group's recommendation was the impetus behind starting the landfill siting process because Bluestem's board listened to the group's recommendations and adopted the system.

The next step was to develop a strategic plan for siting a new landfill. Some of the components of the plan were hiring outside consulting services and defining a coherent process for obtaining public input. The initial strategic plan included offering a benefit to the community or area that would eventually host the landfill. With the strategic plan in hand, we then held a meeting with what we called our "affected parties" (different agencies, departments, and groups that we felt might be contacted during the process), because we wanted to act with all possible foresight. We knew that at a certain point in the process Bluestem would no longer be the main source for information. And we knew that at some point everyone would start questioning what we were doing. We wanted other agencies and county and municipal departments to be aware of the process so they could provide assistance. Having asked for input from the "affected parties," we then initiated the central process.

The first move was to assemble another public input group, a citizens' advisory committee. This time we had about twenty people. Once again, they came from a variety of backgrounds and experiences, and we met with them for 5 or 6 months. The first meetings were mostly educational, providing background on what the focus group before them had done, what materials they had gone through, and what recommendations they had made. We also discussed integrated solid waste management and gave them quite a bit of background on the geology of Linn County and Iowa and on how a modern landfill is built and its various features, including landfill gas, leachate collection systems, and liners.

This new group's mission was not to question whether a landfill was needed, it was to implement the system recommended by the focus group by finding an appropriate location for the landfill. We knew that as this process proceeded, people would start questioning and criticizing. We knew that members of the advisory committee might start thinking, "Well maybe we ought to look at different technologies." We made it very clear at the beginning that reconsidering such alternatives was not part of their mission.

They needed to develop the criteria that would be used to select an appropriate site. There would be three layers of criteria. The first layer, "exclusionary criteria," would exclude areas of the county from further evaluation. The next layer, "delineation criteria," would delineate sites within the area that was left after applying the first layer. Within the "zones of suitability," potential sites would be found. The last layer of criteria would be used to rank potential sites. We wanted to make sure that the advisory committee developed all the criteria before they were applied, because we did not want knowledge of actual site locations to affect how the committee developed further criteria.

Their main concerns at this point remained environmental protection. For instance, one of the exclusionary criteria required a particular amount of depth to bedrock. They wanted to make sure there was plenty of soil between the bottom of the landfill and the top of the bedrock, providing another layer of protection for groundwater. In implementing the exclusionary criteria, they were also interested in protecting certain individual properties in proximity to rivers and streams, and in other environmental issues. For the delineation criteria, they were concerned about proximity to neighbors, proximity to the waste centroid (the weighted center of waste generation in Linn County), and access to roads. The point I am trying to make is that their concerns were environmental in nature: to make sure the landfill would protect the natural environment as much as it could and provide an adequate buffer for neighbors while also being close enough to where the waste was generated.

Once the delineation criteria were applied, thirteen possible sites were identified. This was interesting because the sites were scattered throughout the entire county. We had used geographical information system (GIS) programs to apply the criteria, so we had excellent maps of how much land each criterion excluded (see figure 7.1). The advisory committee also excluded all land that had already been developed. This excluded all the incorporated areas.

Throughout this planning and siting process, we were lucky if two or three people attended the advisory committee meetings, even though

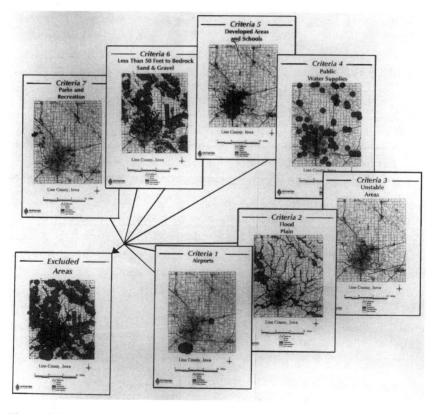

Figure 7.1
No landfills here! Areas excluded from further consideration by the citizens' advisory committee. (Source: Bluestem Solid Waste Agency.)

they were well publicized and coverage by the media had been excellent. It was not until we announced the thirteen sites, at what we refer to as "the meeting in hell," that we started getting huge groups. Prior to the meeting we had contacted the owners of the prospective properties and the media. At the meeting, there were close to four hundred people. The agenda was straightforward: to present the process that led up to the delineation of the proposed sites, present information about the proposed sites, and then allow time for public input. But the day of the meeting was only a week after the property owners had been notified, so the level and intensity of emotions were high. The hardest part of the meeting was listening to the comments of the audience. There were tears, grandstanding, cheering, clapping, and many horrible accusations about Bluestem staff. On a personal level, the worst accusations to hear were those about the lack of concern, and, of course, knowledge that Bluestem staff must have about the environment in order to even propose a landfill be built on their property. Even more difficult was the fact that we were not able to respond to any of their concerns or accusations. We had to just sit there and listen.

After the "meeting in hell" we hosted what we called "open houses." We held three in different parts of the county so that we could be close to most of the potential sites. We thought of these meetings as an alternative to the traditional public hearing where people have to get up in front of a large group of people and speak into a microphone. We thought the traditional format would intimidate a lot of people. At the open houses we set up booths that provided information about Bluestem, the planning process that led up to the list of the proposed sites, the geology of Linn County and the State of Iowa, how modern landfills are built, host-community benefits that could be provided to the neighbors of a new landfill, and specific information about the proposed sites. Bluestem personnel and our consultants staffed each booth. We wanted to provide an atmosphere where people could come in and talk one-on-one with someone familiar with the process.

The open houses were well attended; each of them drew four or five hundred people. They were also quite difficult. There was a lot of emotion, a lot of yelling and crying. Personally, however, they were much more rewarding and successful than the "meeting in hell" because I was able to respond to the concerns and accusations of the audience. I would not quite say I connected with people, but at least I was able to sit down and explain things from my perspective. I could empathize with them. Overall, it seemed like a much more civilized way of presenting information and gaining input.

The Process and the County Government

The whole time we were meeting with the citizens' advisory committee, we were also working with Linn County's director of planning and zoning on an "exclusive use" zoning ordinance. Simply put, the purpose of the ordinance was to provide the county board of supervisors with a checklist they could use when they would be asked to approve a site to be developed as a landfill. We worked with the planning director very closely during the whole process. I had at least ten drafts of the ordinance in my office. Our consultant met with the director several times to make sure the county's ordinance would be consistent with applicable state and federal landfill laws and regulations. We were fully aware of the planning director's process and of the ordinance, and she was very aware of our process; she had even attended several of our citizens' advisory committee meetings. Yet shortly after we announced the thirteen sites, the planning director called my boss, Liz Christiansen, about 10 o'clock the night before the public hearing that would be addressing the ordinance and said that the county was changing the ordinance and we needed to be aware of it.[4]

Language was added to the ordinance stating that landfills could not be built on land with a corn suitability rating (CSR) greater than 65. (A CSR is a commonly used measure of farmland quality; a rating of 65 or higher basically designates land as prime agricultural land.) It had always been part of the county's land use policy to avoid development on that type of land. But development had never been *prohibited* through ordinance, and obviously there had been development on land with a CSR greater than 65. The problem for us was that all thirteen potential landfill sites were located on land with CSRs greater than 65.

All this was happening while we were holding the open houses. After the county board of supervisors adopted the changed ordinance, we met with the citizens' advisory committee to explain what had happened. One of the county supervisors came to the meeting and attempted to explain the ordinance. We also met with our own board; they recommended that the citizens' advisory committee continue with the process, rank the thirteen sites, and recommend two or three sites for approval. We would figure out what we were going to do about the county's new ordinance later. We felt optimistic about getting around the ordinance because when it was presented to the public, the planning administrator announced that the ordinance might be illegal. It might be a form of "exclusionary" zoning rather than "exclusive use" zoning.

When we went back to our citizens' advisory committee, we presented them with the recommendation from our board. But they refused to continue with the process. They said that there was no guarantee that—despite the fact that the ordinance permitted the board of supervisors to waive the CSR requirement—the supervisors would not back out of the process after the sites had been narrowed down to two or three. They did not trust the board of supervisors. They believed that no matter what sites the advisory committee came up with, if a site had a CSR greater than 65, the supervisors would not approve it. Since all the sites were in unincorporated parts of the county, the board of supervisors would make the ultimate decision on the local siting approval process. The committee said that they would not continue with the process; they wanted to start over and establish new criteria. And this time (because of the input that they had received during the open houses and in writing) they wanted to deal only with "willing sellers." A lot of the comments from people attending the open houses had focused on private property rights and the condemnation process and on the fact that many of the properties were farms that had been in their families for years.

In Pursuit of Private Willing Sellers

Following the desire of the citizens' advisory committee, we advertised for willing sellers. The advisory committee made some adjustments to the criteria, and we came up with new maps showing the area that was still available after applying the new exclusionary criteria, excluding all land with a CSR greater than 65. We included the map in an advertisement for willing sellers in the official county newspaper, the *Cedar Rapids Gazette* (see figure 7.2).

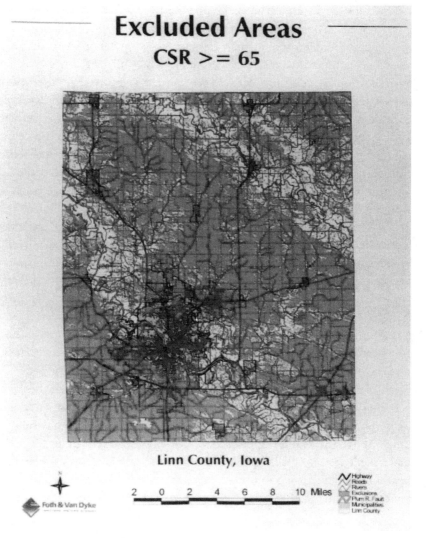

Figure 7.2
Correction: No landfills *here*! Areas excluded after applying the new exclusionary criteria. (Source: Bluestem Solid Waste Agency.)

This whole process of looking for willing sellers took about a year. First, several property owners contacted us and we started negotiating with them. Finally, three properties emerged from this part of the process. Throughout this year, we were conducting all of the negotiations outside of public meetings because we wanted to protect the privacy of the property owners. We wanted them to remain anonymous in case they decided to withdraw their offers. There was no sense in dragging them into a controversial and contentious process and needlessly upsetting their neighbors if they decided not to complete negotiations. But this private part of the process was difficult compared with our open meetings to which the public had always been invited. Now, of course, we received a lot of criticism about everything being done outside of public meetings.

What came out of private negotiations was the Hennessey property, the most appropriate site in terms of the criteria that had been established and the only property where we had made significant progress in negotiations. Having paid for the right to go on the property and allowed to conduct hydrogeological tests, we concluded it was an appropriate site. We met with our citizens' advisory committee and with our board of directors to recommend this site, and the advisory committee concurred that we should pursue it. We announced to the public that we were negotiating with the property owner.

Unfortunately we were unable to reach an agreement with the Hennesseys. They wanted $10 million, but the market appraised value was only about $1.75 million. As a public agency, we are able to pay only so much above the market appraised value—20 percent, I think. Since we had reached a standstill in the negotiation, our board decided to take possession of the property by exercising its power of eminent domain (or condemnation). In that summer, we started the condemnation process.

The Process, the City Government, the State Government, and the Federal Government

The condemnation process took place while the state legislature was considering changing the laws concerning condemnation in the State of Iowa. Because we were uncertain about the result of that deliberation, we wanted to start the condemnation process before the rules were changed.

Also, because the Hennessey property was within the city limits of Cedar Rapids, we had to go to the city council of Cedar Rapids for approval. This is about a 6-month-long process. We did everything that was required: we held the required public hearing plus several more. There were also quite a few public meetings in which we presented the landfill proposal to the Cedar Rapids Planning Commission. In December of that year (1999), the city council approved the local siting request.

That was good, but the state code also says that construction of a landfill must start within 1 year after local siting approval is received. So we had only a year to complete our hydrogeological testing, the cultural resources investigation, and the threatened and endangered species investigation. The last of these alone takes a whole year to do because it has to be done during different seasons. Foth and Van Dyke, a regional consulting firm based in Wisconsin with an office in Des Moines, conducted these investigations required by Iowa law.

We had begun the investigations and were preparing to apply to the state Department of Natural Resources (DNR) for a permit when we first heard about a Federal Aviation Administration (FAA) reauthorization bill. A portion of it dealt with the proximity of landfills to certain types of airports. The legislation would prohibit new landfills from being built within 6 miles of certain types of airports—those that have commercial traffic but mainly smaller aircraft. At that time Congressman Schuster of Pennsylvania was the chair of the committee that doled out money for U.S. Department of Transportation (DOT) highway and airport projects, so he controlled a lot of money and was very influential. He introduced this portion of the bill in order to stop a landfill in his Pennsylvania district. This was going to be a regional landfill that would accept waste from New York City. In other words, he was writing federal legislation to stop a local project. And of course, the proposed landfill at the Hennessey site was almost exactly the same distance from the Eastern Iowa Airport in Linn County as the Pennsylvania landfill was from its local airport!

After we learned of this legislation, our director, the director of the Eastern Iowa Airport, and the mayor of Cedar Rapids went to Washington, D.C. to talk to FAA officials and our local legislators about the proposed law. We were unsure whether the Eastern Iowa Airport would fall within the category of airport that this bill was trying to regulate. We were also unsure about where the measurements of distance would begin and end: the end of the runway, the closest runway to the property boundary on the landfill, or the edge of the property of the airport to where the waste is placed? Another thing we had to find out was how far along in the construction or the development of the landfill one had to be in order for this new law not to apply. We felt that we had done whatever was needed in order to have this law not apply, but the FAA informed us that we were not far enough along in the process to be exempt because we were just beginning the permitting process. We were basically told that the FAA law would stop our project.

Our board decided to halt the condemnation proceedings because they felt we would no longer be able to use the Hennessey site. Then of course we found out from our attorneys that there was no precedent for abandoning a condemnation process this far along! There was no case law that our attorney could use to find out whether we could do it. Our board decided to abandon the condemnation anyway. We did everything that we needed to do to abandon the condemnation.

About a month and a half after our board made that decision, we received a letter from the FAA stating that the law would *not* apply to the Eastern Iowa Airport. We went back to the board and told them of the FAA's letter. To complicate matters even more, earlier in the spring we had negotiated a mediated settlement with the Hennesseys that relied on their higher appraisal. In March, a month before the FAA reauthorization bill became law, we had tentatively settled on a purchase price of $3.75 million. We had begun the mediation process because we discovered that in order to apply for a permit to develop a landfill, we needed clear title to the property or the signature of the titleholder. Before we knew of the FAA complication, we had wanted to settle out of court in order to gain title so that we could apply for the DNR permit.

It Was Hard

I cannot adequately convey how difficult it was to be the spokesperson for the solid waste agency at this time. There were parts of the process I could not talk about because of the private mediation that was in progress. The going back and forth between condemning the property, trying to settle with the property owners, and then all of a sudden deciding to abandon the condemnation was exhausting. To an outside observer who only read articles here and there in the newspaper or heard a snippet on the news, and also to the neighbors of the proposed site, the process must have looked pretty dodgy. It was difficult to explain. Even though everything made complete sense to me, I could not tell all of the story. The board was making difficult decisions at that time. Part of my job as the spokesperson was to protect them while trying to explain what was going on. Also, not having a legal background, it was difficult for me to navigate the whole condemnation and mediation process and then the abandonment of the condemnation. Legally, there were all sorts of wild ramifications and possibilities. It was just hard to explain the situation to everyone and to keep on top of everything so that I could condense information into sound bites for the evening news.

After receiving the letter from FAA exempting the Linn County landfill from the new legislation, our board wanted to make sure they had explored all of their options pertaining to developing the Hennessey site as a landfill. An important remaining issue affecting all options was that state requirement to start construction within a year of local siting approval. Since it was now late summer of 2000, we would have to receive a permit and begin construction by the end of the year. We had to find out if this could actually be done. When we abandoned the condemnation, we had stopped the investigations necessary to gain the permit. We had finished the cultural resources and hydrogeological investigations, but not that for threatened and endangered species. We conferred with the DNR to see if we could be granted a temporary permit and receive the full permit after we finished the threatened and endangered species investigation. The initial discussions went very well, and we felt that we would be able to get a temporary permit so that we could start construction by the end of the year—if our board would decide again to abandon the abandonment of the condemnation process. Try to explain this to the general public and the media! You can imagine what it was like working in the heat of all this.

Then we found out from the DNR that they probably *would not* provide a temporary permit because of a section in the state code which says that one has to establish the *need* for a landfill. We thought that by going through our integrated planning process we had established that need. However, the DNR was questioning that. We were not even sure if the

DNR would approve our needs assessment, even though we had gained the local siting approval from the City of Cedar Rapids. They began asking questions about whether we had investigated transfer hauling the waste to other counties or other states. This amazed me because we had gone through what I thought was by far the most extensive solid waste planning process that had ever been conducted in the State of Iowa. No one had ever gone through a process like this. But also, no one has successfully sited a new landfill in all of Iowa in the 25 years since the state's landfill siting law was written.

When we informed our board of everything that was happening, they decided not to pursue the new landfill. There were just way too many "ifs," way too many things out flapping in the wind, for them to spend any more money to develop a landfill on the Hennessey property.

Given the board's decision not to proceed with the landfill, we had to develop a set of options. We came up with eight and arranged them on a continuum. On one end of the spectrum were options that included the greatest amount of Bluestem involvement; at the other end were those requiring minimal Bluestem involvement. Specifically, the option at one end was to continue with the landfill siting process and at the other it was to disassemble Bluestem and transfer all of the responsibilities for solid waste management back to the separate local governments in Linn County.

At present, one thing is certain and it greatly affects the advantages and disadvantages of the options that the Bluestem board is considering: the local approval awarded by the City of Cedar Rapids to develop that site as a landfill has expired. If Bluestem, through litigation, would end up owning the property and decide to develop it as a landfill, we would have to go through the local siting approval process again. I do not know if the board is willing to do that. Several board members have said there is no way they will ever do that again or pursue that site anymore. Even if we can abandon the condemnation process and get our money back, we still might end up with the property because if the owners have reinvested the money paid to them and cannot pay it back to us, they might just give us a quit claim deed to the property.

Some people have suggested that we could develop the site as a park. Well, yes, that is what everyone wants us to do: donate the land to the Indian Creek Nature Center because it is a beautiful 450-acre wooded site. But its large wooded area is part of the site's appeal as a landfill because it would provide an adequate buffer for its neighbors. We would use only about 80 acres of the 450 as the landfill and would leave the rest in its basically natural state, as a buffer.

One might say this sounds like a familiar NIMBY tale. In fact, we had tried to minimize NIMBY-type conflicts. A lot of people told us, "You will never site a landfill with this process." They said that "site and defend" is the only way any agency is ever going to build a landfill, meaning that it does all the investigating on its own, it comes up with a site, it announces it, it defends the hell out of it, and it ends up in court, which it knows will happen anyway. It just waits and spends all its money at that point. The Bluestem board decided not to do that, and they based their decision on a recommendation from staff that we should not just ask for public input at the end of the process, but actually have the public be a part of the process. I still think it was appropriate to cultivate public input from the very beginning. This was not a bogus public input process. The citizens' advisory committee developed the criteria that were used to find appropriate sites, and the criteria were based on what their concerns were. Their concerns were being a good neighbor and providing an adequate buffer, but most of all protecting the environment.

The Best and Worst of Politics and Planning

This 5- or 6-year experience has been highly negative, but it has also been the most interesting thing I have ever done. It has been extremely difficult because of the controversial nature of the task and because I have been the spokesperson for the solid waste agency. People were constantly saying, "You don't know what you're doing," "You're lying," "You're manipulating, you're abusing your power," and "You're not listening to people." Those kinds of statements have been made over and over again, but I think what we have done has been the exact opposite of that.

I have seen the best of politics and the worst of politics. At worst, elections got in the way of good decisions; ordinances influenced by the private sector were written. However, I have also watched real leaders take action. Whereas some people have questioned the leadership of the county supervisors because of the adoption of their CSR ordinance, I saw some of the city council members of Cedar Rapids really shine and stick with what they felt was the right decision. They had to spend a lot of political capital during an election year. That has been nice to see.

I think I also have developed a much clearer idea of what the role of the planner is, and I really like that role. It is to provide information to the decisionmakers and to keep them on track. This can be a lengthy process, and decisionmakers are busy and have many other interests and issues to deal with. Even though this landfill siting has probably been one of the most difficult decisions the local officials have ever had to make, it is still only one of hundreds that they make every day. Our job is to make sure that they stay focused and that they have the information they need to make a thoughtful and rational decision. As a planner you can be as influential as you want to be. This is difficult. You have to be careful not to let your own views and ethics get too involved because you want your information to remain objective and complete.

Not only are decisionmakers busy with hundreds of decisions, but they are also besieged by the public's negative responses. In this landfill process they were just beaten up at every meeting we had. For the last 4 years we have had people showing up at public meetings without ever making a positive comment. Perhaps two or three times someone has stood up and said, "You're doing the right thing." It is all negative, and keeping the decisionmakers on track has been difficult. A lot of times they wanted to abandon the public input process, saying that it was too hard and cost too much money. I know that deep down they really do not feel that way, so you, as a planner, have to continually remind them why we are doing what we are doing.

Editors' Introduction to Chapter 8

Having written about narrative in planning for a number of years, in this essay Mandelbaum places narrative within a repertoire of support tools that shape policy and planning debates. While narrative and other tools—theories, models, and information systems—are often seen as competitors for attention and credibility, Mandelbaum here presents them as complementary resources. In Planning Support Systems: Integrating Geographic Information Systems, Models, and Visual Tools *(2001), Richard K. Brail and Richard E. Klosterman have created a collection of essays that can provide one context for Mandelbaum's claim. However, they do not include narrative among the tools described in their project. In* Hamlet on the Holodeck *(1997), by contrast, Janet Murray does address the imbrication of narrative and computer technology, which creates virtual and interactive spaces. Arguing that single, linear narratives are no longer acceptable in our contemporary globalized society, Murray offers narratives in virtual reality as the appropriate tools for the current condition. The implications of this latest technological medium for communication—following the rock, the pen, the printing press, the microphone, the radio—intrigue scholars of narrative, who raise questions pertinent to policy study and planning.*

To understand just how important and complex is the question, what stories can be told about and through a computer-generated model, the reader can expand urban boundaries to the full extent of their global implications and trace the contested histories of international efforts to measure and predict regional and global climate change. These efforts, which began more than 15 years ago are highly dependent upon modeling and information systems. They led to the publication and release in January 2001 of Climate Change 2001: The Scientific Basis *(Houghton et al., 2001), a 1,000-page report prepared over 3 years by the 516 contributing authors of the Intergovernmental Panel on Climate Change (IPCC), and to the tentative adoption—but not by the United States—in November 2001 of a treaty implementing the Kyoto Protocol, an international political document underwritten by the scientific evidence in* Climate Change 2001 *and related reports (Revkin, 2001).*

As a planner you can be as influential as you want to be. This is difficult. You have to be careful not to let your own views and ethics get too involved because you want your information to remain objective and complete.

—Michael Berkshire

8

Narrative and Other Tools

Seymour J. Mandelbaum

Planning Support Systems

Planning is such a ubiquitous set of practices that under most circumstances we barely notice how we manage to get from here to there, from aspiration to plan. When, however, our schemes are complex, our uncertainties profound, and our prospective path both obscure and strewn with hazards, then only fools rush in. Prudence dictates careful attention to the ways we construct the future.

In the Academy and in the Field, professional planners—rarely fools—often concentrate their attention by distinguishing between a contested core of planning practices and a set of consensual tools that support those practices. The core is a professional craft, mastered only through a long apprenticeship. The craft varies from one planner or planning unit to another, each acting in a mode that is "in character," but that resists articulation as a general theory or a coherent set of formal principles. Planning appears in the guise of a mystery whose mature votaries cultivate their own styles (Flyvbjerg, 2001).

In contrast, there is no mystery in support systems no matter how technically difficult they seem to novices. The tools can be formally taught and, while practice in the Field makes perfect, the basic skills can be successfully learned within the Academy. The tools are routinely carried from place to place without modification. Properly employed, they yield the same results for every practitioner.

The distinction between planning processes and the systems that support them is an imperfect heuristic device. The idiosyncratic craft of planning necessarily softens and complicates the hard edge of the tools with which it is coupled; the aura of mystery (and unique insight) that surrounds the idiosyncratic craft is inevitably penetrated by the consensual claims of the support tools that are accessible to anyone with the time and patience to master them.

The analytical difficulties in distinguishing where planning begins and support ends highlight the intellectual richness of the overtly simple notion of Planning Support Systems. What sorts of systems will support what sorts of planning? What do we mean by "support"? Does support depend on a critical distance from core planning processes or an intimate engagement that obscures intellectual boundaries and organizational distinctions?

"Planning Support Systems," as a term of art in the company of professional planners, characteristically describes computers, information systems, and both mathematical and iconic simulations.[1] It is, however, clear, at least in my eyes, that it would be impossible to construct plans as accounts of the future if we did not craft our memories into manageable narratives. It is not equally clear how narratives discipline our construction of plans; how and when narrative gives way to other tools.[2]

I have struggled in the past with the relations between the construction of pasts and futures, and I am not eager here to rehearse familiar arguments about the forms and limits of both conservative and apocalyptic visions. I offer instead a simple observation: the interorganizational networks that shape policy and planning debates command a repertoire of support tools. The four principal elements in that repertoire—narrative, theory, model, and information system—often appear as stylistic competitors in a race for intellectual dominance: narrative will liberate us from the abstractions of theory, geographic information systems from the simplified dynamics of mathematical models, and models from the authoritative reading of history as destiny.

I prefer, however, to see the repertoire as a set of complementary resources.[3] I start with the "other tools"—theory, model, and information system—before turning to the forms and uses of narrative.

The "Other" Tools

A theory is a set of literary forms rather than a single protocol. When we invest ourselves in those forms—when we "have a theory" of something or other, including planning itself—we are (rightly or wrongly) encouraged to believe that we can make sense of our worlds even if we cannot control or repair them. Theories reassure us by converting boundless particular assertions into a small number of general claims, and then embolden us to explain and justify those claims parsimoniously. We watch a dance and infer a culture; justify a difficult choice and imply an entire ethical order.

A theory is useful as a support system because of its nomothetic bias—its focus on things in general. That same quality is its greatest weakness. Theories—whether of planning or of the phenomena being planned—illuminate pervasive structures, but cannot "completely" explain the choices of particular agents in particular settings. Theories and theorists fail if they foolishly attempt to satisfy demands for "full" explanations and "complete" information. Even if they resist those seductive demands, theorists and theories are repeatedly frustrated by agents playing a game against them by acting out of character or in an unanticipated frame; for example, drivers trying to beat the expected pattern of rush hour congestion, rule-breaking stock market traders, imaginative litigators, and creative negotiators.

Models, like theories, come in many forms. They appear as physical structures, system diagrams, field trials, matrices, equations, communal myths, metaphors, and institution-defining rules. In somewhat different ways, all of these forms allow those who build models to understand how streams of behavior are sustained and modified, how both "mistakes" and "successes" are engaged and disengaged. A theorizing sensibility encourages us to grapple with particulars by treating them as members of a general class. Models, in contrast, place particulars within the flows of a system—say, my personality—in which a small set of relatively stable elements can support a large number of different and protean states.

The ability of a model to simulate behavior lies, paradoxically, in our willingness to accept a dissimulation; to trust in a similarity in abstract relationships without the security of identities of form and the appearance of reality. We read evocative metaphors as figurative rather than literal texts. We see a skyscraper in an architect's foot-high model, a road network in the lines of a computer program. Even a field trial of a new welfare policy requires that we accept a rarefied miniature, insulated from the ordinary slings and arrows of political life, as if it represented the "real thing."

The powerful specificity of simulations and the ambitions of their builders ensure that models will sometimes be confused with the worlds they mimic, endowing them with more authority than they deserve. Paradoxically, that same confusion also ensures that some users (with or without the prodding of the builders) will obscure the distinction between models and the planning processes they support. Models are often used to help the parties in difficult negotiations cultivate a shared understanding of their contested choices: "Let us agree—if we can—that this is how the building will look or how the tax reduction will radiate through the economy." In that difference-reducing role, models do not replace the ordinary crafts of reciprocity and compromise. Without those crafts, the tools of simulation will fail.

Models are sometimes employed in the early stages of planning when conventional assumptions and metaphors are challenged and divergent paths explored. That early use is not the ordinary staging of simulation. Both formal symbolic and iconic models characteristically enter into planning processes when the builders believe that they broadly understand systemic relations and have narrowed the set of problem definitions and feasible interventions. That belief and the use of simulation to narrow the domain of the possible lends an authoritative cast to modeling that often challenges the judgment of political leaders. The examples of intellectual authority asserted (as in National Academy of Sciences consensus statements) and challenged ("just a model") are legion: They include models that simulate global climate changes, the allocation of tax burdens, the impact of transportation choices on air pollution, and the path of the national economy. Consider, as an illustration of such assertions and challenges, a characteristically strong statement by the former Mayor of New York City, Rudolph Giuliani. In December 2000 he dismissed academic critics of his stadium proposals with the following:

I think they're very wrong. I don't think they've ever run a city. I don't think they've ever run anything. They're doing some kind of economic model. (*New York Times*, December 22, 2000, p. 22)

Information systems—the last of my "other" tools—organize measures of the attributes and performance of the worlds we inhabit. Wrapped around and through the entire planning process, information systems discipline policy debates, defining what it is possible to say in ways that will be understood across the field.

Information systems are no less politically contentious than theories and models. Think of the recent debates over statistical adjustments and racial categories in the census of 2000 or over the measurement of academic achievement as a central element in education policy making. Nor are information systems technically simple. Measurement, representation, and the recognition and assessment of patterns present conceptual difficulties that are entangled with—and equal to—those associated with modeling. An invalid or unreliable measure of output confounds attempts to simulate the dynamics of even a simple system. New global planning institutions struggling to represent systems that are large, complex, and varied are fated to multiply mistakes. When there are so many ways to be honestly deceptive, the practices of lying with statistics are difficult to distinguish.

Despite those similarities between information systems and models, there is a significant difference in their political cast. Information systems that support planning are not composed of theoretical explanations or simulations, but of loosely coupled fact-claims. Adding new claims or new organizational protocols to an information system may be expensive (particularly in comparison with free data), and debates over the purported benefits of knowledge investments may mask deep political conflicts. New facts do not, however, characteristically require the conceptual realignment associated with the integration of innovative models or theories into support systems. We commission studies or add items to census schedules in response to practical and politically significant queries. (The national bookkeeper attends closely to the client.) The uncertainties that generate those queries may surface at every planning stage. They appear in the early diagnosis of problems and in the shaping stages of projects, plans, and policies when divergent thinking is welcome and constraints are not yet embedded in political alliances. They appear again when alternatives are prospectively assessed and then when plans are implemented and we ask retrospectively whether our efforts have made any difference.

The overt (even if deceptive) simplicity of fact-claims makes information systems seem much more politically accessible than the abstract and hierarchical forms of theories and models. In the culture of professional planning, information systems and their technologies—the New Information Technologies (NITs)—dominate the conception of planning support. (The capitalization here signals their importance and announces a mythic history of the "knowledge society.") NITs frame a new challenge to a democratic planning ethos. How can publics participate in planning if they cannot access the Internet and interpret the spatial representation of information, if they cannot track a set of contingent "what if" statements on a digitized map?

Narrative

The narrative craft is so widely shared that it may seem strange to think of it as a technology. It is not only universal, it is soft. It belongs to the language of humanists, not engineers or scientists. Yes and no! There is a soft or at least politically contentious side to the other tools—theories, models, and information systems—and symmetrically, a technology to narrative in support of planning.

At the simplest level of folk wisdom, we all know that there are great storytellers and some so clumsy that they could not tell a simple joke properly if their lives depended on it. At a more formal level, narrative is constructed by a series of three linked protocols that may be learned even by those of us who lack the storyteller's knack. The first of these sorts the elements of a story into foreground and background. The narrative that serves the purposes of the author is set entirely in the foreground. The background may change over the course of the narrative and may influence the people and events to whom the author asks us to attend. It does not, however, claim our imaginative engagement and understanding. Authors and readers (or narrators and listeners) may of course differ in their purposes and their instrumental calculations. As a result, they may be inconstant (or simply different) in the ways they sort the elements of the story they are reconstructing, creatively moving people and events from one setting to the other in the hope of negotiating a shared understanding. If, however, the parties to the construction of the narrative fail to distinguish foreground and background, purposes are unlimited and stories are out of control. (Only God can tell such a tale.)

The second protocol prescribes that narratives, unlike life itself, must have a beginning and an end. We do not need to know about events before the beginning in order to grasp the purposes of the narrative. (In the world of Genesis we do not need to know what God did before the creation of heaven and earth.) Events after the end of the narrative are similarly irrelevant if they do not add to the cogency of the account.

This second protocol of beginnings and endings connects the construction of historical narratives to the creation of plans. There are some narratives that make no sense until we create a new beginning—all the old beginnings are deeply flawed—or are emotionally or ethically incomplete until we bring them to a new conclusion. In both circumstances, narrative gives way to action and history to planning. This second protocol does not, of course, ensure consensual judgments. It only delineates the reasons that authors and readers may argue and the terms within which they may come to agree.

The third protocol guides authors and narrators in endowing individuals and social entities with recognizable identities, grounding the interactions of the protagonists in their characters, memories, and aspirations. It creates the principal contents of the narrative form. The protocol encodes the narrative in one of a small set of dramatic myths and imputes existential choices to the collective mind of groups and communities. The narrative form and its contents are so compelling that without them experience seems chaotic; with them, we invest even the sparsest narratives with a verisimilitude that endows them with the authority of a realism that other support systems cannot match.

The history of fields of knowledge is frequently told as a developmental journey from story and fact to theory and model. That is not the tale told here. None of the four tools described in this essay disappears in the development of planning. None is a merely persistent primitive that we retain for reasons that defy our modernist logic. Instead, the tools complement one another as equal and interrelated partners.

Editors' Introduction to Chapter 9

Planning Director Karin Franklin tells what happened when her city began envisioning a new mixed-use neighborhood. It would be more walkable and diverse than conventional subdivisions; it would even contain a town square that would act as a focal point and a place for neighborhood gatherings. Constituting a quest for community, and seeking to evoke a strong sense of place, her story and the project it describes sit squarely in the middle of recent debates about the desirability and feasibility of New Urbanist designs. On one side are New Urbanists who advocate major changes in neighborhood and regional design; they seek diversified rather than homogeneous neighborhood populations and they want to facilitate far greater use of public transportation. On the other side are those who seek to make new neighborhoods more walkable and neighborly, but who often end up designing—as in Seaside, Florida— cutsey towns for wealthy people. Some are on neither side. As John A. Dutton documents in New American Urbanism *(2000), even the second set of New Urbanists find their projects being opposed by nearby property owners who do not like increased density. A different set of priorities, evident, for example, in Throgmorton's argument in this volume about tenticular radiations, raises the question of whether either variant of the New Urbanist vision by itself would significantly alter the health of America and the environmental burden of its wasteful economy.*

Ironically, the Peninsula would return New Urbanism to the midwestern small-town architecture and ideology that spawned it (see figure 2.1 in chapter 2). What does this return mean for sustainability in a context where the old neighborhoods themselves are being altered by practices of economic development that transform the older architecture and neighborhood values? Is it a return with a difference?

Drawing upon Eckstein's concerns for the fate of community stories in the production of a single plan, we can ask of this narrative what happened between the articulation of community views (in the vision task force and the design charrette) and adoption of the Peninsula design plan? In other words, can the discursive democracy Beauregard advocates emerge in the context of the representative democracy that structures Franklin's planning practice? Can ideas generated in discursive processes retain significant influence when the interests of propertied elites constrain locally elected representatives?

It is, however, clear, at least in my eyes, that it would be impossible to construct plans as accounts of the future if we did not craft our memories into manageable narratives. It is not equally clear how narratives discipline our construction of plans; how and when narrative gives way to other tools.

—Seymour Mandelbaum

9

The Peninsula

Karin Franklin

The story of the Peninsula project, as it has become known, started with small groups of Iowa City residents, separately but together, trying to find a way to regain a sense of community and neighborliness in the places they lived and the city we were building. Iowa City, home to the University of Iowa, was the envy of many small towns in Iowa at the end of this past century. It was growing at a moderate rate compared with many cities in the country, but it was booming by Iowa standards. With growth and the evolution of the community from a midwestern college town of 33,443 in 1960 to a small city of 59,738 by 1990, many people felt the loss of community and neighborhood that often accompanies such change. Traffic was increasing. You had to drive to get to any shopping area. The 15-minute "rush hour" changed to a real rush hour as people experienced longer commutes to work. Lines got longer at stores and there were many more unfamiliar faces.

In 1992, the city staff initiated a new comprehensive planning process in which a number of topical task forces were created to address the vision the citizens of Iowa City had for its future beyond 2000. This was not a particularly startling or innovative initiative, but it was a departure from how Iowa City had approached comprehensive planning in the immediate past. Instead of developing a plan at the staff level and then bringing that plan to public hearings, which few people attended, the idea was to start with at least some of the people and develop a plan based upon what residents thought they wanted for the future of the town.

Although there were nine different task forces that met separately, a number of them gravitated toward the idea that we, as a community, had to build our city differently than we had in the past. We needed to focus on the people who lived here. We needed to build neighborhoods, not subdivisions. People should be able to travel relatively easily to see neighbors and shop and go to work. The street should look as if people lived on it, not cars. Diversity in neighborhoods was important—different types of households, different income levels, different types of housing, even other land uses than residential. Open spaces ought to be available in each neighborhood for parks or places where people could get together on occasion if they wanted to. This all sounded like the New Urbanism we, as professional planners, had been reading about in the planning literature.

New Urbanism, or neotraditional planning, is trendy right now. For that reason, the terms are used by many people and are often associated with a certain dogma. There are purists who insist that all the details and nuances of New Urbanism—the mix of housing, corner stores, and places to work in the neighborhood—must be met in a project, and if they are not met, the project has failed. Similarly, there are those who will reject any project associated with New Urbanism in any way, feeling that it smacks of social engineering by government bureaucrats and other rose-tinted nostalgists. The valuable thing for Iowa City was that many of the principles of New Urbanism were what the residents who participated in the Beyond 2000 task forces seemed to be seeking. The trick for the staff, the planning commission, and the city council was to determine what features of New Urbanism fit our community and where and how. Would a shift in how the neighborhoods of the city were built be accepted by the local development community?

The opportunity for an experiment came in 1993 with a devastating and persistent flood. Iowa City was established as the territorial capital in 1839. The city straddles the Iowa River, a tributary of the Mississippi, with the downtown just up the hill from the river, and the "guts" of the city—the sewage treatment and water plant—relegated to the river's banks. The 1993 flood drove home the imperative of moving the city's 100-year-old water plant out of the Iowa River floodplain and the foolhardiness of building housing in flood-prone areas. As a consequence of extensive flooding in the Mississippi River valley that year, the Federal Emergency Management Agency made money available to cities to purchase floodplain land and take it out of the inventory of developable land forever. Iowa City capitalized on this opportunity by buying some privately owned land in the Iowa River floodplain in which there also happened to be alluvial wells that would supply a relocated water treatment plant. However, the property for sale also included approximately 90 acres of uplands that the owner would not separate from the floodplain property (see figure 9.1).

In 1995, the city council decided to dip into the city's general fund to purchase the 90 acres of upland ground in order to acquire the needed well sites and enable acquisition of the federally funded floodplain. It was understood at the time that the 90 acres would eventually be sold, with the revenues from the sale used to replenish the often-stressed general fund. There was some discussion of retaining the upland as open space, but even the most environmentally conscious council members agreed this was financially difficult. The question then became to whom should the land be sold—just anyone willing to pay market value or should the city use this opportunity to try something different?

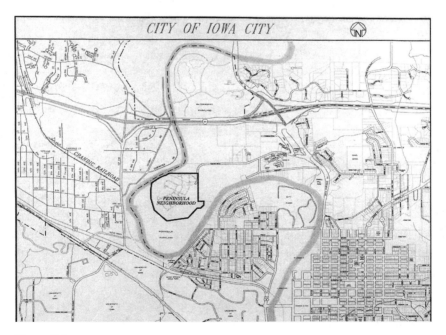

Figure 9.1
Location of the Peninsula neighborhood. (Source: Iowa City Department of Planning and Community Development.)

Shortly thereafter in 1997, the senior planner for the city, Robert Miklo, attended a conference sponsored by the National Homebuilders Association in Beaufort, South Carolina. He saw the work of Vince Graham there and met Victor Dover of Dover, Kohl & Partners, a land-planning firm that follows the principles of New Urbanism and the work of Andres Duany and Elizabeth Plater-Zyberk.

After sending out requests for proposals to various firms, Iowa City hired Dover, Kohl & Partners in 1998. Their task was to work with the city planning staff, the council, the planning and zoning commission, and interested residents in developing a plan for how the community would like to see the 90 acres—the Peninsula—develop. A full-day charrette was held on a Saturday in a church sanctuary, the largest space available close to the actual property. The people in attendance were a mix of neighborhood activists, developers, historic preservation advocates, architects and designers, lower income housing providers, citizens interested in a new place to live, and just the curious. Councilors, planning commissioners, and city planning staff were also mixed in the group. Victor Dover, a charismatic, comfortable person who worked very well with the expected midwestern skepticism about any outside consultant from South Miami, started the day with a brief history of development patterns in the United States and what this New Urbanism thing was about. Then everyone was able to take a bus and go the short distance (about 2 minutes) to the property to be developed.

The site is one of the most beautiful in Iowa City. Upon entering the Peninsula, one is on an upland plateau bordered by indigenous oaks and hickory trees. At the edge of the plateau, a break in the trees reveals the lower-level fields and riverine woodlands that define the land formed by the bend in the Iowa River. This was truly a site for a city, or at least a neighborhood, on a hill. Having a sense of the place, everyone boarded the buses, returned to the church for lunch, and broke into small groups to draw the neighborhood they thought should be built in this place. The day ended with the small groups reporting to the whole assembly, with many similarities among the proposals.

The next week the consultant team, including designers and architects, worked in the Iowa City Council chambers, inviting people to come in and share ideas, comment on the plans being drawn, or join in on coloring the perspective of the neighborhood being developed. The outcome

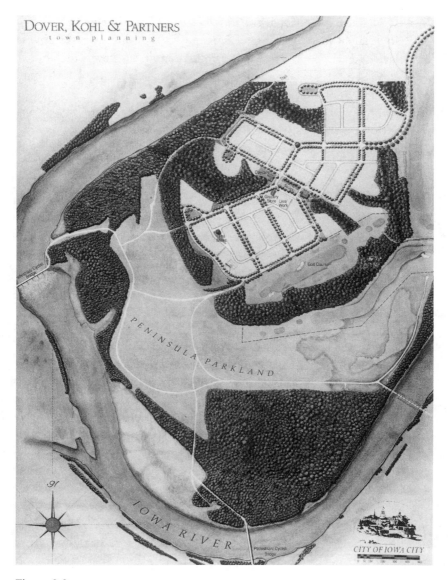

Figure 9.2
Site plan for the Peninsula. (Source: Iowa City Department of Planning and Community Development.)

of this collaborative effort is the development plan shown in figure 9.2. The city council formally adopted this plan and authorized recruitment of a developer in 1998. The vote was unanimous and even the council was surprised at the unanimity and enthusiasm with which they endorsed this approach.

A number of principles became central to what the community, as represented by the council, wished to achieve in development of this area. This neighborhood is intended to embody the goals of the Beyond 2000 comprehensive plan and include a diversity of housing types and thereby a diversity of household types and income levels. To ensure that the neighborhood would be accessible to at least some people in the lower income levels, provisions in the request for proposals from developers required the successful development team to address the issue of long-term affordability. How this would be done was unknown at the time. It was well known that historically, New Urbanist neighborhoods were pricey and increased in value over time if they were done well. In Iowa City, housing affordability is already a challenge, with the median price of a new single-family house being $133,000 in 2000. The desire of the council was to try to build a neighborhood that made possible a healthy mix of income levels, from at least 80 percent of median income ($50,300 for a family of four in 2001) on up, at the start of the project and as it matured.

To engage a developer to build this neighborhood, requests for proposals were sent out in November 1998. The responses were not good in terms of numbers. Of three anticipated proposers, two backed out because of financial and workload factors within their companies, and we were left with only one proposal to review. Although there were many positive things about the proposal, the full team of the proposer was not on board with the concept of the neighborhood. The issue of affordable housing for lower-income households was dealt with by a proposal that the developer give money to local housing agencies, but not provide any space in the development for more affordable units. The decision was made by the staff in consultation with the city council to go back out to see if we could obtain additional proposals. Three different entities came back with proposals and T. L. Stamper, L.L.C. was chosen in March 2000 based on their team's experience in New

Urbanist design and development and their commitment to the Dover, Kohl plan.

In the developer's agreement made between the city and the selected development team, 10 percent of the total number of housing units will be made available to local nonprofit housing corporations or the Iowa City Housing Authority. Ten percent of the cost of the land will be discounted for these agencies to provide housing for low- to moderate-income households. This is to ensure long-term affordability through the transfer of property to agencies focused on the mission of affordable housing. In addition, the mix of housing types provided allows the entry of varying income levels as the neighborhood is built. Theoretically, this mix will enable access by varying income levels as time goes on. The housing types include row houses, town houses, cottages, bungalows, apartments, and estate houses.

Another strong goal and principle of this project was to build a neighborhood in which the visual image projected focused on people—people with feet. The street presence is that of sidewalks and porches; garages are set back or oriented to a rear alley; streets are narrow. The living space of people dominates; cars are a tool, not the primary occupants of a dwelling (figure 9.3).

Access to open spaces is also important. Although the actual neighborhood is urban and appears dense in comparison with more conventional subdivisions, access to open spaces is abundant and prevalent. The use of a single-loaded street[1] to skirt most of the neighborhood ensures that access to the lowland park of more than a hundred acres that rims the neighborhood is not blocked by private property and that the vistas of this area and of the river valley may be enjoyed by all. Within the neighborhood, a town square provides a focal point and a place for impromptu or planned get-togethers (see figure 9.4).

It was recognized from the beginning that this neighborhood would not embody all the parts of what New Urbanist neighborhoods *should* have. The site is not large enough, nor does it have the access, nor will it have the traffic required to support any significant commercial development. Live-work units[2] will be a possibility. In time and with a population of young families, a day-care facility may be built. A small, very specialized restaurant may succeed . . . if it is very good and very

Source: Dover, Kohl
& Partners and
Ferrell Rutherford
Associates, L.L.C.

Figure 9.3
Design illustration of a streetscape for the Peninsula. (Source: Iowa City Department of Planning and Community Development.)

expensive. However, a retail establishment or a grocery store is unlikely. So for the purists, the Peninsula will not truly be a New Urbanist community. I do not believe this is important to the city council or to most people in Iowa City. What is important is that we have the opportunity to recapture some of the neighborliness of our old neighborhoods and that there is more choice in the kinds of new neighborhoods we can live in.

At this point the zoning of the property is in place and the preliminary plat of the entire neighborhood has been approved by the city council. Now that we are almost beyond the question of whether this project will be built or not, there is some speculation in the community about whether it will succeed. There are those who steadfastly maintain that this is the kind of new development people have been seeking for years. Many of these people are the housing consumers. Others, primarily local developers and financial institutions, question whether people in Iowa will want to live so close to one another on small lots.

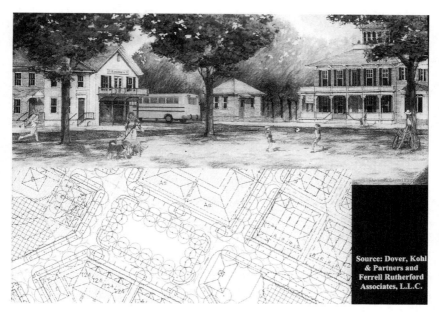

Source: Dover, Kohl & Partners and Ferrell Rutherford Associates, L.L.C.

Figure 9.4
Design illustration of a town square for the Peninsula. (Source: Iowa City Department of Planning and Community Development.)

The success of this project will no doubt be measured in different ways, depending upon one's perspective. If lots sell at a steady rate, buildings are constructed in a timely fashion, and the project is completely built in 5–8 years and fully occupied in 10 years, most people will deem it a success. If property values increase over time, the owners will consider it a success.

However, if all of these things happen and the initial public purposes in undertaking the Peninsula project are not met, this effort will not be deemed a success by many of us involved. One of our goals was to try to build a place in which, through good design, different building types could be juxtaposed so that varying income levels would have access to the neighborhood. If the property appreciates to the extent that the neighborhood becomes financially inaccessible at the lower ranges of the housing market, it will not meet one of the goals of the city in developing this project. Given the initial availability of lots to nonprofit housing corporations, there should always be units available to the 50–80 percent median income level, surpassing the council's original goal. However, the question remains of whether those units on the "open" market will become so expensive as to be inaccessible to many people in the moderate income levels.

If no one among our local developers builds a similar project, this will also be seen by many of us as a failure of this project. One of our goals was to show the local development community, who had little to no experience or desire to take the risk in building these types of neighborhoods, that there was another way to build new parts of the city. Through this project we hoped to seize the opportunity to provide a model of a different way to build new neighborhoods. If no one copies this model, some of our effort will have been for naught.

My belief (hope) is that the Peninsula project will be a success in many ways and that others in Iowa City or in the area will build similar neighborhoods that will also succeed. That through the success of these neighborhoods, a new way to approach development will evolve, offering another choice to housing consumers, and that these neighborhoods will embrace a diversity of people, households, housing types, income levels, and land uses. Time will tell.[3]

Editors' Introduction to Chapter 10

Acknowledging that a wide array of academic disciplines are more engaged than ever with the spatiality of human life, geographer-planner Soja argues that geography remains the excluded sibling within planning, especially planning theory, and within U.S. universities more generally. He asks us to consider the power, indeed the hegemony, of storytelling as the vehicle of history and advocates an interpretive practice that disrupts the linearity of history with the simultaneity of geography understood on multiple scales—from the urban to the regional to the global. Sustainability, we argue throughout this book, depends upon the apprehension of just this simultaneity.

Not simply dismissive of story, Soja here tells his own planner-practitioner narrative, but in a disruptive form that deliberately and unsentimentally engages in a self-interested stakeholder's complaint, like those that planners routinely hear, if not tell. Soja places this common mode of complaint in an uncommonly spatialized form that makes explicit use of the more transformative elements of story that Eckstein explicates. In doing so, this essay foils the reader's expectation of Soja's usual theoretical discourse and attempts instead to use storytelling to take the reader by surprise. Such a defamilarizing narrative may drive home his repeated theoretical arguments if more conventional theoretical discourse has not.

Artifice-conscious narrative and spatially conscious narrative interpretation have practical implications in the consideration of land use and the built environment. Architecture scholar Christine Boyer (1994) turns to innovative literature in pursuit of theoretical assumptions and aesthetic forms that can bring coherence to the historical fragments of overweening historic preservation and postmodern architecture. More generally, any stories and theories of storytelling that serve to remind audiences—planners and stakeholders among them—that time and space are mutually constituted serve to remind them as well that any local land use project must be seen in its proximate and distant, past, present, and future contexts. Such reminders are also central to environmentalists' vision of sustainable places, which require simultaneous local consciousness of both the regional and global impact of local everyday practices, those tenticular radiations Lawrence Buell defines in Writing for an Endangered World *(2001).*

Time will tell.

—Karin Franklin

10

Tales of a Geographer-Planner

Edward W. Soja

What can possibly be wrong with storytelling? Stories are minihistories and in history we can find not only the vast storehouse of accumulated knowledge about what has gone before but also vital ties to the immediate present and the imminent future. Stories are our most ancient form of social communication, a way of creating and maintaining community and culture through the sustaining voice of past experience. They can inspire, entertain, inform, comfort, teach. What, then, is there to dislike about a good story, especially when it is told with enthusiastic conviction and the best of intentions?

In her theme-setting essay "Making Space," Barbara Eckstein initiates a discussion of some of the potential problems involved in storytelling in the particular context of current debates in planning theory and the creation of sustainable cities. As she points out, persuasively told stories can carry with them an extraordinary self-empowerment of the storyteller as well as an excessive emphasis on the status quo against the challenges of change. My intent here is to broaden and deepen our understanding of the hazards inherent in the "storytelling turn" in urban and regional planning through some very personal stories about being a scholar who is simultaneously a planner and a geographer. Being a geographer-planner, as I will try to demonstrate persuasively, is the source and provocation of my concern over the more hidden hazards involved in storytelling.

I begin with a general conclusion: the practice of persuasive storytelling must be approached with caution, not because storytelling and the narrative form more generally are not attractive and powerful ways of understanding the world, but because they may be too powerful and compelling, silencing alternative modes of critical thinking and interpretation, especially with regard to the spatiality of time. What follows is elaboration.

Storytelling Is No Longer What It Used to Be

Caution and care are particularly necessary when telling stories today, in an era of globalization, economic restructuring, and instantaneous communication. Simply stated, storytelling is no longer what it used to be. It has taken on different dimensions after what has been a long period of extraordinary change at every scale of human life, from the concretely local to the encompassingly global. With increasingly thickening layers of the new and different blanketing contemporary life, the past becomes more distanced from the present, simultaneity takes precedence over sequence, the synchronic demands more attention than the diachronic. Here are some epigrammatic signposts that help us understand this restructuring of storytelling and its attendant narrative form.

We are at a moment, I believe, when our experience of the world is less that of a long life developing through time than that of a network that connects points and intersects with its own skein. (Foucault, 1986, p. 22)

It is scarcely any longer possible to tell a straight story sequentially unfolding in time. And this is because we are too aware of what is continually traversing the story line laterally. (Berger, 1974, p. 40)

How, then, can I translate into words the limitless Aleph? . . . Really, what I want to do is impossible. . . . What my eyes beheld was simultaneous, but what I shall now write down will be successive, because language is successive. (Borges, 1971, p. 13)

Many story lines are intertwined in these critical observations, each drawing us into lively contemporary debates, not just in planning and urban studies, but in nearly every academic and professional discipline. These discursive debates involve the relations between, and the ways we think about

space, time, and social being
geography, history, and society
the production of space (Lefebvre), the making of
history (Marx), and the constitution of society (Giddens)

It is becoming clearer that the components of each of these triads operate on equal terms, each constructed by and constructing the others, forming a triple dialectic. Yet there is a powerful tradition, a persuasive story line, that persistently lingers to tip the balance toward an implicitly closed dialectic between history and society, with the production of space and the critical salience of human geographies left muted and peripheral, always there but never at the forefront. The continuation of this all-encompassing history-society dualism reduces critical spatial thinking to a mere addendum, even though attention to the spatiality of human life has become more widespread today than it has been at any time in the past two centuries.

This now less apparent but still formidable hegemony of the historical-sociological imagination, or what might even be called a kind of socialist historicism, throws a cautionary light on too enthusiastic an embrace of storytelling and the narrative form, even when it is advocated by spatially sensitive scholars. So much has accumulated over the past two centuries to load the intellectual dice against the spatial or geographical imagination, inside as well as outside the field of urban planning, that thinking as a geographer-planner makes me continue to be concerned about the ease with which storytelling can too readily assimilate and subsume critical spatial thinking in its wake.

My concern begins with the very act of writing. The relations between the production of space, the making of history, and the constitution of society are enmeshed in simultaneity, but this enmeshing must be translated into a successive text, a written narrative. As Borges laments, language itself makes this necessary. What I have to say will therefore be constrained by its own obligatory narrative form, for every text unavoidably tells a story, induces a historical reading. Nevertheless, I will simultaneously try to do what I can to break open the narrative form to alternative writings and readings, creating as much room as possible for an active and critical spatialization.

Planning as Friendly Persuasion

The particular tales I tell spring from what can be described as the postmodernization of planning theory. The field of planning has traditionally been seen by its most progressive theorists and historians as a vehicle for enlightened social change, a practice aimed at improving society, especially for its most disadvantaged. In comparison with architecture and many other fields, a powerful postmodern critique came relatively late to planning, and when it hit, it was met with an unusual combination of forceful resistance, uncomfortable attraction, and utter confusion. As was the case elsewhere, long-standing modernist theorizations, especially those that positioned the planner as public hero (few heroines were allowed in these theorizations, a significant part of the problem), began to crumble. But the planner, as among the last proponents of progressive modernism, found it more difficult than most to let go of many of its inherent attachments.

Amidst this heightened ambivalence and in reaction to a fear of numbing paralysis, a new planning theory has begun to emerge, cautiously selecting relatively safe ground within, or at least near, the sprawling borderlands between modernism and postmodernism. Few planning theorists have been willing to make an explicit choice between the two turfs, instead preferring some seemingly neutral and comfortable mediating zone where the best theoretical practices of the past can continue to thrive in updated guises and with new forms of professional legitimation. It can be argued, I believe, that the revival of storytelling in planning emerges primarily from this relatively safe ground, finding reassurance in acts of friendly persuasion while being protected from having to deal too directly with most of the bewildering complexities of the postmodern world.

Once . . . upon a *time* . . . the modern story continues. Here are selections from the front page of the November-December 2000 issue of *ACSP Update*, the newsletter of the Association of Collegiate Schools of Planning.

Telling Stories in School:
All about Planning Practice
One of the most challenging problems in teaching planning, especially to master's students, is linking course content to real-world examples. Planning theorists have highlighted the importance of using stories—both in the practice of planning and in the pedagogy of planning education. Even though every successful planning professor has a stock of stories—examples [stories equal examples?]—that are consistently effective in communicating to students, we have too few opportunities to share our stories with other planning faculty.

Here is your chance to share your stories without having to spend too much time and effort in the process. . . . Submit your stories of up to 5000 words (shorter is acceptable, too) to *ACSP Update*.

Time for a quiz:

1. What city has epitomized low density sprawl and unsustainability in the planning literature and in the popular press?

2. What major metropolitan region has the densest urbanized area in the U.S.?

3. In what major metropolitan region have suburbs become the most compact and dense?

4. Using traditional measures of density and sprawl, what is the most sustainable urbanized area in the U.S.?

The answer to all four questions is Los Angeles.

Just last night my wife, Maureen, said to me while reacting to some pretentious "reality" program on television, "Doesn't storytelling mean that they are making things up? Why are they so pleased to say they are telling stories? Isn't it like telling lies?" I refrain from explaining that there are other definitions of storytelling.

Storytelling (in a nutshell) is

Part of a grand tradition of oral history
A means of defining community
An effective teaching tool for planners
Like case studies in business or law schools
A powerful alternative to scientific analysis
A kind of professional psychotherapy
A form of communicative action, following Habermas
A compelling notion in planning theory (see above)
A reflection on planning practice
A way of constructing the future
A means of persuasion
A way of manipulating time and place
The very act of plan making and plotting

O God! I could be bounded in a nutshell, and count myself a King of infinite space . . . (*Hamlet*, II, 2; first prescript to Borges, 1971)

The Plot Thickens

A mouse click on my computer's thesaurus brings up a fascinating chain of meanings for "plot": connive, conspire, contrive, devise, plan, scheme, chart, diagram, graph, map out, arrange, concoct, tabulate, collude, machinate, manipulate, map, sketch, design. Take your choice.

The plot I am thickening here has to do with the relations between time and space, the historical and geographical imaginations, and in particular being a historian versus being a geographer. More specific issues arise from these relations. What does it take to be called a planner? A planning theorist? Why is it easier to understand what a planning historian does than what a planning geographer or geographer-planner might do? Who is more likely to be involved in critical theory and thinking, a historian or a geographer?

One thing is clear. There is something familiar, comfortable, in attaching history to planning, and especially to the intellectually charged subfield of planning theory. This is true not just in planning but also in neighboring fields such as architecture and urban design, where there are long-established bonds among theory, history, and criticism. However, space and spatial thinking play a particularly central role in planning, architecture, and urban design, in their theories and practices. Why then is there no explicit mention of space (or geography) in the bundling of theory-history-criticism? And why is it so difficult for a geographer, like myself, to be called a planner or a planning theorist, even after nearly 30 years of teaching in a planning department?

Perhaps space is so familiar to planners that it is taken for granted, implying that we are all spatial planners? Or is this neglect of the spatial coming from a (unexamined) bias that intrinsically favors and privileges history and the historical imagination over geography and the geographical imagination, especially with respect to theory and critical thinking?

As is usual in these moments, I turn to Foucault.

Did it start with Bergson or before? Space was treated as the dead, the fixed, the undialectical, the immobile. Time on the contrary was richness, fecundity, life, dialectic. (Foucault, 1980, p. 70)

The great obsession of the nineteenth century was, as we know, history: with its themes of development and of suspension, of crisis and cycle, themes of the ever-accumulating past, with its preponderance of dead men and the menacing glaciation of the world. . . . The present epoch will perhaps be above all the epoch of space. We are in the epoch of simultaneity: we are in the epoch of juxtaposition, the epoch of the near and far, of the side-by-side, of the dispersed. We are at a moment, I believe, when our experience of the world is less that of a long life developing through time than that of a network that connects points and intersects with its own skein. . . . Ideological conflicts animating present-day polemics oppose the pious descendants of time and the determined inhabitants of space. (Foucault, 1986, p. 22)

What a story these words begin to tell. An enchainment of events and inclinations that runs through a century and a half to the present, mapping an epoch in which history—a decidedly historical if not historicist narrative—framed, articulated, and critically intervened in our scholarly and theoretical (but not necessarily practical) understanding of the world. History (and mainly his-story) dominated theory and critical thought everywhere, history-theory-criticism became unbreakably molded together, leaving little room for the "other."

Against this historical and modernist grain, Foucault confidently asserts the imminence of a spatial turn, the birth of a new (postmodern?) epoch that shifts attention from the ever-accumulating past, from history telling and a strict sequential narrative, to the networks of simultaneity and juxtaposition that define the conspicuous spatial dimensions of the present-day world. We are now above all in the epoch of space, he says.

But isn't his assessment premature, to say the least? Even with the extraordinary transdisciplinary diffusion of critical spatial thinking that has taken place in the past 10 years, isn't the critical historical imagination still dominant, still hegemonic? Isn't it still working to constrain critical spatial understanding so that even a spatial discipline like urban planning could remain tightly in the grips of its disciplined history? By turning to storytelling in planning, are we engaging (unknowingly?) in an ideological conflict between those with a critical geographical imagination and the "pious descendants of time" Foucault talks about?

Is this why I am so hesitant to accept the abundant virtues of storytelling and feel so strangely combative and out of place?

More Theoretical Asides

I have been telling the same story, over and over again, for more than 20 years, constantly searching for better ways to convince and persuade. It is not direct disputation that drives me to repetitive elaboration, but rather a tacit if not complimentary acceptance of the persuasive story that persistently misses the point. . . .

My opening foray was an essay on what I called "the sociospatial dialectic" (Soja, 1980). I argued that social relations and spatial relations, society and space, need to be seen as mutually constitutive; that contrary to the prevailing Marxist view, the social is not inherently favored over the spatial. In other words, the sociality and spatiality of human life are simultaneously intertwined and interdependent. And they are often at odds with one another, opening up a multitude of story lines that have rarely been followed by social scientists or scientific socialists, because the possibility of such balanced yet problematic interdependence had never been seen as a serious issue.

In *Postmodern Geographies* (1989), I built a similar argument about a spatiotemporal dialectic, about the interdependent intertwining of space and time, geography and history, the spatiality and historicality of human life. Just as an exaggerated socialism and social scientism favored the social over the spatial in the first dialectical pairing, blocking recognition and understanding, a persuasive historicism, the implantation of causal interpretation and explanation within an essentially temporal narrative and critical historiography, prevented a more balanced appreciation of the relations between space and time. This variant of historicism, linked to a related sociological imagination, tended to submerge and subsume interpretation and explanation drawn from critical spatial thinking and a decidedly geographical imagination.

In *Thirdspace* (1996), I connected the two arguments in what I chose to call a "trialectic," an ontological tying together of human spatiality, sociality, and historicality on equal terms, with no one of the three privileged a priori over the other two. If this ontological premise were accepted, I thought, then all else would logically follow. However, significant barriers remained in achieving such a three-way balance, especially given the relative weakness and narrower scope of the spatial and geographical versus the historical and sociological imaginations. Even today, no one is startled by the claim that everything in human life is, always has been, and will always be intrinsically historical and social; and further, that our understanding of everything is enhanced by critically examining this existential historicality and sociality. Saying the same for spatiality is another matter.

What became clear in the 1990s was that while there was growing recognition of the critical importance of geography and spatial thinking, there remained a conceptual barrier that prevented the historical and the geographical imaginations from reaching an interpretive and analytical equivalence. History continued to be seen as more encompassing and incisive, even within the so-called spatial disciplines. Until the scope of the geographical imagination could be expanded to the same all-embracing existential level that the historical and sociological imaginations had already achieved, there could be no three-way balancing of critical perspectives. This expanded scope, I argued, could be reached through what

I called a "thirdspace perspective," akin to what Lefebvre (1991) defined as lived space and Foucault described in "Of Other Spaces" (1986) as heterotopology.

Merely proclaiming this interpretive equivalence and arguing at an ontological and epistemological level about the relative neglect of spatiality with respect to the historical and the social was not enough. Some more empirical demonstration of the critical power of this expanded (third) spatial perspective was needed. In *Postmetropolis* (2000) I tried to deliver these empirical goods by taking a new look at contemporary urbanism and exploring how emphasizing a broadened spatial perspective might elicit innovative new insights into our historical understanding of the (uneven) development of human societies, starting with the old debates on the origins of cities and stretching to present-day discussions of globalization and economic restructuring.

How persuasive these spatial stories have been is still difficult to tell. But to some degree my persuasiveness no longer matters very much, for something quite remarkable began to happen in the late 1990s, well beyond any influence I may have had, I hasten to add. A critical spatial perspective was being creatively adopted in a wider range of academic disciplines than ever before, in art history and music, in cultural studies and ethnography, in film theory and communications studies, in literary criticism and poetry, in economics and accounting. In what may in retrospect turn out to be one of the most important intellectual developments to emerge in the late twentieth century, a growing number of scholars began to interpret the spatiality of human life with the same critical insight and analytical power that has traditionally been given to life's historicality and sociality.

The point of these theoretical diversions: those promoting the usefulness of storytelling and the narrative form in planning and elsewhere need to be constantly aware of the challenges raised by the critical rebalancing that is taking place among historical, geographical, and social modes of analysis and interpretation. Given the dual hegemony of social and historical thought over the past century and a half, this rebalancing act primarily involves a stronger and more explicit emphasis on critical spatial interpretation, on effectively spatializing the sociological and historical imaginations.

Such spatializing efforts, however, are not just a matter of adding geographic information or a bevy of spatial metaphors to one's analysis or narrative. Spatialization is necessarily a deeper and more disruptive process, intended to raise an explicit consciousness about the degree to which conventional modes of analysis and interpretation, and especially the narrative form, in often very subtle ways muffle the strength of critical spatial thinking.

It is not enough then simply to recognize the importance and usefulness of the spatial dimension, to sympathetically "make space" for it in social thought and storytelling. One must work hard to reconstitute interpretation (and storytelling) in new and different ways.

Stated in another way:
Storytelling is primarily about the historicality and sociality of human life.
Historicality is explored primarily through the narrative form.
Sociality is explored primarily by attaching social thought and theory to the narrative.
Storytelling in planning depends on traditional forms of narrative historiography.
Storytelling in planning draws very selectively from social theory.
The social theory most heavily influencing storytelling in planning is from Habermas.
Storytelling in planning theory is thus primarily narration as communicative action.
Habermas is probably the least spatial of major twentieth-century social theorists.
Dealing with space becomes a major problem for planning as storytelling.
Even when persuasive planners have the best of intentions.

Being a Geographer-Planner—Part I

Does writing a history or a biography always involve storytelling? Is there anything that separates the terms and concepts of history, narrative, story? What about a fable, a chronicle, a novel, a film, a lie? Is writing a history or a biography also always spatial? Isn't all life, everyone's lived experience, simultaneously spatial, temporal, and social? Does the concept of a spatial biography make you uncomfortable?

In almost every English-speaking country in the world except the United States, there is a close institutional relation between geography and planning. For example, at the London School of Economics, where I have recently been teaching one term each year, regional and city planning is a specialized course in the Department of Geography. Somewhat similar arrangements exist at major universities elsewhere in Great Britain and in Canada, Australia, New Zealand, India, and South Africa. And again in great contrast to the U.S. norm, the field of geography is a relatively popular subject, and geographers from these countries are disproportionately represented on the faculties of the best American geography departments. The same, however, cannot be said of the best American planning schools, where there are relatively few geographers of any kind.

Despite a recent upsurge of national interest, instigated in part by the growing popularity and widespread use of geographic information systems (GIS), geography in the American educational system (and in the general education of most Americans) is a weakly developed discipline. There are currently no geography departments at any Ivy League university (except for a small survivor at Dartmouth) or at such prestigious private universities as Stanford, Massachusetts Institute of Technology, and the University of Chicago, although nearly all of these universities were at one time flourishing centers of geography education. The recent closure of the Department of Geography at the University of Chicago (a committee remains) is particularly telling because this department was once exceptionally prominent and also had close ties with planning. Such spatially oriented planning scholars as John Friedmann (more on him later) and Janet Abu-Lughod are products of these close ties at Chicago.

Is there something exceptional that persists in the American educational system that devalues geography and keeps geography and planning apart? I once approached one of America's leading theorists of history and the historical narrative and asked him about the persistence of a bias that tended to submerge the importance of geography, to make it decidedly peripheral to historiography, just as it has been peripheral in the academic division of labor at American universities. He responded with surprise. "But I love maps!" he declared. That ended the conversation.

When I was in primary and secondary school in New York City, geography was hardly taught at all, and when it was, it was usually taught by historians rather than by people trained in geography. I read somewhere once that in the period between the two world wars, when a new social studies curriculum was being developed in U.S. schools, geographers were hesitant to contribute, fearing dilution of their established position and neglect for the key role of physical geography. It was also said that others could contribute whatever geography was needed in social studies. Did this begin the marginalization of geography and geographers in America, the reduction of geographic knowledge to unchallenging lists of facts, capital cities, and capes and bays?

When I had my first job teaching in the now defunct Department of Geography at Northwestern University, I remember the chair telling me that he had just received another irate phone call from a parent, complaining that his offspring had been lured, almost as if it were by a religious cult, into becoming a geography major. This was outrageous, the father claimed. He was not paying good money for his son (or daughter) to become, what, a geographer? "I didn't even know geography was taught at the university," he added. Who was to blame for this misunderstanding?

It seems as if I was always a geographer. I called myself one when I was ten, even before I knew you could make a living doing it. When asked at school about what I wanted to be when I grew up, though, I never said "geographer" because no one else in New York City seemed to know what I meant. I usually said "psychologist" or "engineer" or "artist," copying my friends. I went to Stuyvesant High School, finding to my delight that it offered cartography classes as an alternative to woodshop or metalworking. Then, instead of going to the more prestigious City College of New York (CCNY) in Manhattan, which had no geography department and still does not, I went to what was then called Hunter (now Lehman) College in the Bronx, a short bus ride from my home. For the first time in my life I took courses in geography.

Being a Geographer-Planner—Part II

How many planners does it take to change a lightbulb? It depends on how you define planning.

My life as a geographer-planner began in 1972, when I arrived to take up a position in what was then the Urban Planning Program in the School of Architecture and Urban Planning at the University of California-Los Angeles (UCLA). I remember staying for a week at a palm-fronded hotel on Wilshire Boulevard, where my 6-year old son found a friend by the swimming pool who turned out to be the now-deceased English comedian, Marty Feldman. We have some old photographs to help us remember.

I had some serious concerns about making the move from Northwestern University, where I taught for 7 years in geography and in African studies, both of which were then among the best departments in the country. I was tenured by the time I reached thirty and had a fairly rapidly growing reputation as a political geographer and Africanist. Changing disciplines would be difficult, but I was attracted to UCLA for several reasons, the most prominent being John Friedmann, whose work I had become aware of some years earlier and whom I came to know while serving on the committee on comparative urban studies he had helped initiate at the Social Science Research Council in New York. John enthusiastically welcomed my interests in geography and spatial analysis and assured me that these interests would fit in particularly well with the new (Chicago-influenced) urban planning program, established just 2 years earlier. I found out later that Brian Berry, an Englishman then at the University of Chicago and one of the best-known geographers in the country, was the first choice to fill the spatial slot, but he chose to remain at Chicago.

I did not know then how exceptional John and the UCLA program were—and would be for at least the next two decades, an extraordinary period when the Graduate School of Architecture and Urban Planning (GSAUP) was one of the most exciting and productive places in the country for interdisciplinary research and creativity, especially with regard to the intersection of all the spatial disciplines: urban planning, architecture and urban design, geography, environmental studies, and urban sociology.

Although I am frequently described as an urbanist, I have always been more of a regionalist. In my earliest research in political geography and African development geography, and in my education as a traditional geographer combined with an eager purveying of the new quantitative and theoretical revolution in geography, I always thought of myself as contributing to a new kind of regional geography. At UCLA, I fit comfortably in regional planning, then centered in a specialization called "urban and regional development," which also contained studies of urban and rural development in the Third World. I called my regional development courses "spatial planning." But I also taught courses on the political economy of urbanization and, for a while, on urban design and the built environment.

It did not take long before I started thinking that hey, all of planning is spatial planning, that everyone in the program was dealing in one way (or scale) or another with trying to shape the social production of space, from the local to the global levels. We were all engaged in what I called, and began to write about, as "spatial praxis," adopting and spatializing a term (praxis) that featured prominently in John Friedmann's required seminars in planning theory, which he defined as the transformation of knowledge into action. However, my spatializations were kept well out of the planning theory (dis)course. This was privileged ground, demanding a certain purity of consciousness. Putting geography first, ahead of planning, was not acceptable. While I changed my income tax forms to identify myself as a geographer-planner by profession, it was not always easy to be seen as such by the others in the most liberal, open-minded, and spatially oriented urban planning department in the country.

Planning theory, in particular, remained forbidden territory for the geographers in urban planning at UCLA, even when their number increased to at least four by the late 1980s. The two required doctoral courses in planning theory prepared the students, many of whom had little or no planning knowledge or experience, for the capstone colloquium in planning theory, a thematic seminar. This sequence of courses served to inculcate a distinctive identity and ideology among both faculty and students, an identity and ideology that clearly defined and separated planning from its nearest neighbors.

Within the GSAUP, the major separation was between planning and architecture. Although closely tied together historically in both the United States and Europe, the upstart urban planning (at least in the United States and especially at UCLA) had broken off from its architectural roots in the last half of the twentieth century and become more oriented to the social sciences and to progressive political practice. Planning theory worked to rationalize this separation of cultures and traditions. Even in the zone where the two disciplines came closest together, a clear distinction was made. What architects called "urban design" we planners called "the built environment," shifting emphasis away from physical form to questions of community economic development and the politics of housing. Occasionally one of our built environment planners would complain that the planning theory courses neglected their theoretical and professional heritage, but little was ever done to address this.

Geography, housed in the College of Arts and Sciences, was urban planning's major external competitor. Even though GSAUP was a separate professional school, the urban planning doctoral program saw itself as producing (practice-oriented) scholars of equal or higher quality than those turned out by geography or the social science departments. Since nearly all dissertations in urban planning dealt in one way or another with spatial and geographic issues and perspectives, the need to draw boundaries between urban planning and geography, open though they may appear to be, was essential. To the planning theorist, to be called a geographer-planner or to a lesser degree a spatial planner, was either not quite politically correct or else unnecessarily redundant. Such subidentities, especially when they were so close by, had to be kept in the background. The barricades against architecture and geography, each seen as potential "space invaders" seeking to gobble up the true turf of planning, may have been invisible, but they were nonetheless formidable.

Looking Back

As Barbara Eckstein notes, storytelling obtains all the richness of the great African griots, the revered oral historians whose words held societies together; and like the wondrous Australian aboriginal dream lines, storytelling allows time to be harmoniously synchronized with space and place.[1] But she also cautions planners. Stories must be told and read with a careful and critical eye, lest they be obliviously romanced into sentimentality or wielded with such blind authority·that they are used to win people over to believe everything the storyteller has to say. Eckstein writes: "Planners . . . who tell persuasive stories to win over listeners without acknowledging who or what authorizes their individual or collective authorship might well be perceived by audiences as marketing agents producing advertisements." There can be an unbearable heaviness to being a storyteller in planning.

And there is more to be forewarned about. Planners must always tell multiple stories, leave sufficient space for the stories of others to be heard, cultivate a veritable "garden of forking paths" (à la Borges), blooming with a postmodern multitude of "contingent fragments" (quoting Beauregard). Eckstein immerses all this in a sensitivity to just how easily questions of space recede behind questions of time. She thus appeals to storytellers to become "geohistorical," to see how space and time must become equal interpretive partners. Is this way of "making space" for storytelling enough to appease my doubts? Yes and no. Is it at all possible for storytellers in planning to take heed of all these warnings and still tell useful stories? The question remains open, for to close it now is to miss the point.

Let me sum up again by taking a few more forking paths.

In many ways, visual art is the opposite of storytelling. The artist tries to capture meaning without spoken words and usually without written words. However, there are interesting other connections. Eckstein makes one of these links when she turns to artistic practices as a means of counteracting that "certain formal coherence" that often prevents narration, especially in the particular form of "the plan," from capturing the real and imagined complexities of everyday life and events. The "best art," she notes, "defamiliarizes the everyday, encouraging its audience to rethink their humanity and their place in society." Can what is accomplished by the visual be incorporated into verbal storytelling? Can the rigidities built into the narrative form be artfully overcome, or must the narrative be smashed, radically deconstructed, in order to make it more open to the visual and the spatial, as Walter Benjamin once argued with regard to art history and criticism? (cited in Soja, 1996, p. 173).

Flashback 1: Soon after *Postmodern Geographies* was published in 1989, I began receiving rather positive comments from art historians. Since not that many positive comments were flowing in from those in other disciplines, I asked an art historian I knew why there was such a favorable reaction from his field. He explained that in the period between the two world wars art historians and critics began discussing the constraints imposed by the narrative form and conventional historiography on the interpretation of art. In particular, it was thought that such constraints made it more difficult to appreciate the specific power of the visual and related notions of spatial form and meaning. Many concluded that art history had to be cautiously reconfigured to find better ways to incorporate the visual and, by association, the spatial. They did not formally theorize these issues nor were they set in the broader geohistorical context of the development of western ideas about space and time. That I had begun to do this theorization and contextual setting was what probably generated the positive comments. I might add that I have never yet received a single positive comment from historians not specializing in art.

Flashback 2: I recently gave a plenary lecture to a conference of the International Association of Art Critics on the theme of "The City as a Vehicle for Visual Representation." The conference took place at the Tate Modern Museum on the burgeoning South Bank of London. I began my lecture with a critical appreciation of the site itself, part of the globalized resuscitation of what had for two thousand years been London's impoverished periphery and was now glistening with the new symbols of "Cool Britannia." I will not go on with my site description except to say that it was what many in the audience expected of me. The world of the visual arts—in theory, practice, history, and criticism—has recently become deeply engaged with the urban and the spatial.

After my talk, I was invited by a writer for a local art magazine, who seemed to be very familiar with the current literature in critical human geography, to meet for an interview with the director of exhibitions at the Tate Modern (Walker, 2001). To my delight, the director knew of my work and we spoke about a rather playful observation I made in *Postmodern Geographies* that there should not only be art historians but also art geographers, mentioning, for example, the insightful work of John Berger. In 1989, the notion of an art geographer seemed incomprehensible, perhaps even silly. In 2000, it was almost common currency. I was even asked by the director to suggest the names of some art geographers for the Tate Modern to hire to help shape their new exhibitions and to tie the museum more closely to the surrounding urban milieu.

Flashback 3: Moving back to John Berger, I close my story with further quotations from his splendid work, *The Look of Things*.

We hear a lot about the crisis of the modern novel. What this involves, funda-
mentally, is a change in the *mode of narration*. It is scarcely any longer possible
to tell a straight story sequentially unfolding in time. And this is because we are
too aware of what is continually traversing the story line *laterally*. That is to say,
instead of being aware of a point as an infinitely small part of a straight line, we
are aware of it as an infinitely small part of an infinite number of lines, as the centre of a star of lines. Such awareness is the result of our constantly having to take
into account the *simultaneity and extension* of events and possibilities. (Berger,
1974, p. 40)

Every sentence here has a bearing on planning as storytelling. Older
ways of telling a story are no longer as effective as they once were. As
Foucault also noted, the linear sequence is constantly disrupted, broken
down, redirected by what is happening elsewhere, by transversal connections to other stories. Plots no longer simply thicken, they burst open in
many different directions, taking us away from the story line into what
Foucault called "other spaces" and Berger describes as the "simultaneity
and extension" of events and possibilities. A new mode of narration is
developing to meet these contemporary challenges, one that is thoroughly spatialized. To find out how and why this is happening, let us
continue with Berger's prescient observations. Writing, amazingly
enough, in the early 1970s, he sets his explanation in what today would
be called the "globalization process."

There are so many reasons why this should be so: the range of modern means of
communication: the scale of modern power: the degree of personal political
responsibility that must be accepted for events all over the world: the fact that
the world has become indivisible: the unevenness of economic development
within that world: the scale of the exploitation. All these play a part. [And here is
the punch line.] *Prophesy now involves a geographical rather than historical projection; it is space not time that hides consequences from us.* To prophesy today
it is only necessary to know men [and women] as they are throughout the whole
world in all their inequality. Any contemporary narrative which ignores the
urgency of this dimension is incomplete and acquires the oversimplified character
of a fable. (Berger, 1974, p. 40; emphases and brackets added)

There is no need to summarize further.

Editors' Introduction to Chapter 11

More often than not, citizens asked to consider questions relevant to the sustainability of their neighborhoods and their city focus the discussion on crime. As a private investigator constructing the life stories of indigent clients facing death sentences, Barthel provides a view of America's urban criminals, police culture, and middle-class attitudes about crime from the perspective of its poorest neighborhoods and citizens on trial for their lives. In the process of his narration, Barthel notes how the location of these poorest neighborhoods is changing within the larger territory of the city-region.

While for many audiences his is an unusual vantage point from which to view urban geography and its stories, others may know the issues at stake not only through their own experiences but also through the work of scholars and storytellers such as H. Bruce Franklin (1998), James Baldwin (1965), Angela Y. Davis (1971, 2001), Gerald Frug (1999), John Edgar Wideman (1984), and Manning Marable (1991, 2000). Since Davis and Barthel share the same home territory of the San Francisco Bay Area, she may offer the most immediate context for his essay. After her early 1970s autobiographical work on prisons and prison reform, she returned to the topic with force in the 1990s. Interested readers can track much of that extensive body of recent work, including a substantial 1998 conference at the University of California–Berkeley and published venues as diverse as Social Justice *and* Essence, *through the website for the organization Critical Resistance.[1]*

Planning scholars, who are not always so attentive to the place of crime and attitudes about crime in their forecasts, models, and designs, can turn to any and all of these writers for an analysis complementing Barthel's. Together this body of work provides intersections of understanding about the prison–industrial complex, prison literature, crime, borders, middle-class attitudes, and city building that planners can use to frame their practice.

Storytelling (in a nutshell) is
Part of a grand tradition of oral history
A means of defining community
An effective teaching tool for planners
Like case studies in business or law schools
A powerful alternative to scientific analysis
A kind of professional psychotherapy
A form of communicative action, following Habermas
A compelling notion in planning theory (see above)
A reflection on planning practice
A way of constructing the future
A means of persuasion
A way of manipulating time and place
The very act of plan making and plotting

—Edward W. Soja

11

The Meanest Streets

Joe Barthel

Since my occupation is rarely represented in academic circles, and is perhaps curious, I think I should introduce myself. I am by profession a private investigator (P.I.). This is perhaps the most mediated profession in the world, and it is nearly impossible to extricate the reality of what I do from the media haze behind which it is obscured. Investigators are ourselves complicit in the mythologizing. There are few of us who do not have a movie poster on the wall or try occasionally to sound like Bogart. I myself have a Maltese falcon behind my desk. At an oral history conference, an earlier foray back into academic life, I was able to break through some preconceptions about PIs by coining the term "forensic life historian." Whatever it is called, my job has given me access to the stories of individuals and communities at a depth and immersion that few, if any, other occupations afford.

I focus primarily on two areas. The first is death penalty cases, in which I work for attorneys representing indigent defendants. I spend hundreds of hours over a year or two delving into the lives of men (usually, but not always, men) either facing trial for their lives, or already condemned to await execution on Death Rows around the country. (As Mumia Abu Jamal has observed, Death Rows are the fastest-growing public housing projects in the United States.) In these cases, I try to dig out the documentary history—the intersection of the private lives with public records—and to identify, locate, and work with the storytellers who can help explain the pressures on a person's life that can lead to an explosion of murderous violence. Jurors most want to know, literally, "What's the story here?" And if we can begin to answer that satisfactorily, can find and present stories and storytellers to whom they can relate and from whom they can begin to understand the defendant's life, they can make decisions out of compassion rather than fear.

These are multigenerational explorations, bound only by time and imagination. (Time translates into budgets. There are few politicians these days who will stand behind adequate budgets for the representation of indigent defendants.) We try to situate a defendant/appellant within the personal, family, community, and broader social dynamics that affect their lives. I—and a few others doing similar work—trace the sources of rage.

It is rare, though not unheard of, that anyone welcomes me into their lives or eagerly relates the tales I need to hear—of multigenerational physical and sexual abuse, of undiagnosed congenital mental heath problems, of economic hardship and desperation, of neglect and maltreatment by social institutions that nominally exist to help them, of humiliation by bosses and neighbors. That is, the stories of lifetimes of frustration. Often family members, at least initially, would rather risk the death of a loved one than relate the family embarrassments that might save them. However, the stories do come out, and usually people become grateful that they have. And the storytellers usually, with help, develop the strength to walk into that most embarrassing and humiliating venue, the courtroom, and tell their stories.

And jurors generally listen. In interviews with scores of jurors, many of them have told me—groping for the language of their own stories—that jury duty has been the most morally profound moment of their lives, the one time when they have been forced to listen to stories of people whom they have habitually passed by and ignored; the one time they have been forced to get behind the horror stories that they consume daily on front pages, tabloids, and televison; the one time they have had to look at the subdivisions and fissures in their "communities"; the one time they have had to make conscious and deeply consequential moral choices.

The other cases I specialize in are civil rights cases. Here I work on behalf of plaintiffs who think they have been wronged: women denied promotion, gays and lesbians who felt harassed in their workplaces, African Americans who didn't get the house or the service to which they felt entitled, Latinos and Filipinos who are tired of the fact that every time they speak the language of their childhood, white supervisors are sure it can only be to "dis'" them. In these investigations I am also asked to analyze the broad social forces at play in an encounter that itself lasted only seconds, but in these cases I may have only a few hours, not months, to do it.

I describe this practice at perhaps more length than I should because it is the foundation, the unusual vantage point, I bring to discussions of cities and storytelling. I am not going to attempt a broad synthesis or a deep analytical exposition. I am going to try to tell a few stories about how I got to this position and what I see from it.

As a student some decades ago at Columbia University, I worked in East Harlem part time for a couple years at a settlement house, trying to figure out what I could offer to troubled teenagers just a few years younger than myself. One night a student and I walked through Harlem for hours, talking deep into the night. I got lost, and asked, "How do I get back to Columbia?" His incredulous reply was, "Look up." I did so.

From above, looking down from Morningside Heights, Morningside Park had always seemed to stretch out invitingly. The streets below seemed teeming and full of life. It was a lovely cityscape. From below, as I looked up at night, the park collapsed into an undifferentiated sheet of dark wall, indefinable and unscalable. And above that wall we could see, literally, the glittering chandeliers of the Columbia faculty club. How you see a city is a matter of perspective. This was one of my first experiences with the discipline of *looking up*.

A couple years later, I was doing tenant organizing. To overcome the handicap of being rejected as a distant figure coming in once in a while from outside, I moved into a dilapidated and disease-ridden building run by a major slumlord, in a neighborhood we wanted to organize. There I met an older black man (Robert) who worked as a building superintendent for his lodging, who told me his story. Years earlier he had been a foreman in a warehouse job, with a wife and two children. A friend of his got into a fight one night and was arrested. Because he was a respected neighborhood figure, Robert was asked to go down to the station house and find out what had happened to his friend and neighbor. He did so. At the police station, cops first refused to speak with him, then insulted him, and they argued. Ultimately they beat him, Robert told me, and then threw him in a cell. With no charges filed, he thought they were just preparing a report or something before releasing him. The shift changed. Cops came and went. After a while Robert was transferred to another jail, the cops on the new shift refusing to listen to his protests. A couple of days later he was taken to court, where lawyers shuffled papers for a while, then said "There's been some kind of mix-up here, we've got to find his paperwork," and he was shuttled back to jail. He then was passed to a couple of other holding areas and later to a psych ward for being intractable and delusional in his insistence that he had done nothing and had never been arrested.

Some years later, Robert was taken into an office, told by someone that a horrible mistake had been committed, and he was free to go. He had never had a trial. One day he was snatched and the doors closed behind him, and years later the doors opened just as abruptly. Of course, his wife and children were long gone, his mother had died, the place he had worked had closed. He never located his family again.

I didn't believe the story. It still sounds fantastic in the retelling. To confront my skepticism, Robert lifted his mattress and pulled out some yellowed newspaper clippings kept there between pieces of cardboard. There was his story. It included, as he had not, references to the fact that he had been offered $30,000, I think it was, by the City of New York as compensation for his abuse. Robert and his attorney wanted more. While his case was being negotiated, Robert drifted into drinking and once angrily swung at a cop who approached him, leading of course to an assault charge. The city then withdrew any discussion of further compensation. They lowered the offer of $30,000, and he refused a settlement because it was too little for the loss of his family and his life's platform. His lawyer dropped him because of "attitude." Later, he received $60 for his story from a magazine.

As a young activist I tried unsuccessfully to find the lawyer who had handled his case, and then went to the legal defense fund and public legal agencies to see if it could be reopened. I met a lawyer, a former assistant district attorney, who listened patiently to my retelling of the story and then replied, in tones I will never forget: "Young man, I appreciate your interest and your energy. But you don't understand. This happens every day."

I mention all this because this most important discipline—*looking up*—and this most important insight—unthinkable things happen to people around us every day—probably predisposed me to finding my way years later to investigation and certainly shape the way I approach it.

I tell the stories at length because I have learned that *only* stories have the power to move. So I propose to discuss some stories, individual and collective. Since in my professional capacity, I usually, but not always, deal with poor people, I will confine these remarks to what I may have learned about the stories of poor folks and from my interaction with them.

It is a truism that cities are undergoing turbulent change. In the San Francisco Bay Area it feels that the 50 or so years of white flight to the suburbs, and abandonment of the inner cities to Blacks and Latinos, is being reversed with a fierce vengeance. White money is coming back to town, fast, and driving blacks and Latinos and Southeast Asians and Pacific Islanders and some others to the suburbs, which are beginning to notice expanding pockets of poverty among the suburban sprawl. It is quite frequent now that if I need to find the relatives of a defendant or witnesses to a crime that happened just a year or so ago in Oakland, California, I have to search for them in the streets and jails of Vallejo, Santa Rosa, Fremont, Salinas, or other midsized cities an hour or two away from where they lived just a short while ago.

And in West Berkeley or expanding sections of Oakland, I find either abandoned housing, or "fixer uppers." On weekends there are smiling, industrious, white people installing bars over windows and placing some kind of barbed plants—"guard plants" we call them—around their lawns, while across the street and down at the corner are throngs of unemployed black youth sullenly watching the proceedings. (While such a reaction is often experienced and driven in cities as a dynamic of race, it is also a class phenomenon. In northern California or Central Valley towns where DOT COM Gen Xers have cashed their hyperinflated stock options for downpayments on places an hour or so from their new industrial parks, I often find similar dynamics. However, there it is poor and unemployed whites who fling their empty 40s at the sport utility vehicles in the driveways.[1])

What are the stories that emerge repeatedly from my clients and how might they help us understand what is going on in cities these days?

Mobility and Danger: Contested Space

One cluster of stories concerns the question of scale, distance, and contested space. In my own life, I am aware of how my work has expanded my radius of frequent activity. My daily routine is no longer just a journey to a campus or worksite, a coffee shop or market, a movie and restaurant district, and back home. I enter multiple communities, work regionally and travel constantly up and down the state, and at least monthly I am on planes to some out-of-state place, usually a slum in or near a city of a size similar to my own.

People whose stories I hear may have experienced evictions or moved occasionally from town to town, but in their childhoods, adolescence, and early adulthoods they rarely had an active world that extended beyond a few blocks, often not even that. Alliances are often measured and enforced in units of blocks, even different ends of blocks, and beyond this the turf turns hostile. There are *hundreds* of teenage sets within the larger gang structure in Los Angeles, often with their own signs and symbols, rituals and rites, and mechanisms for enforcing loyalty, block by block. Teenagers and young adults are often nervous, tense, and hypervigilant when they are away from their home ground, even for just hours. The belligerence, brittleness, and *attitude* that middle-class people often detect in poor kids (and whites in blacks) is often fear and anxiety.

In developing social histories I always ask, "Walking out your door at age 6/12/16/22 what did you see? Turn left and walk me a few blocks. Turn right and do the same." Often there is embarrassment at such a silly question, but then the stories roll, and they are often about challenge, fear, and violence.

Just a few months ago, a client in his late twenties told me how he had been picked up by police, driven across town, and dropped off in a strange neighborhood. A plainclothes cop shouted out to the neighborhood, "Hey, a guy from *East* is out here," and drove away. A crowd gathered, and my client had to run like hell to avoid a beating, or worse.

This geography-based anxiety is not confined to the poorest and youngest. Consider how the phrase "driving while black" entered the language, or look at the anxieties related by students and professionals, black and Latino, in Lee Mun Wah's film, *The Color of Fear*. In this documentary of a weekend discussion about racial conflict, Black and Latino men from Oakland at one point describe their nervousness when traveling on "foreign turf" in white northern California; the white man to whom they are talking simply refuses to believe them. *He* has never seen racism there. They become furious because both the realities of their anxieties and their efforts to discuss them are dismissed.

Transportation and Crime

Related to the question of geography, mobility, and safety is another cluster of stories that I frequently hear among those who have been driven to violence. These are stories about automobile problems as major factors in pushing people over the emotional edge. When you or I have a car problem, we are inconvenienced, sometimes seriously. But, we generally can get a loaner, or a rental car, or get by with just one car in the household for a couple weeks instead of two or three. We call the automobile association, have the car towed to a garage, and slap the repairs on a credit card.

When poor people lose their cars, the fragile structures of their daily lives often collapse, frequently imploding into violence. If the man takes the car, the woman may not be able to get to her minimum wage job. Or they can't deliver the kids to the grandmother who cares for them. Or they have to choose between a radiator repair and food. Or they get a couple of tickets for an "abandoned" car or wrong side of the street parking, and the tickets are 20 percent or more of their weekly income, and there goes the birthday present or the Sunday meal. Or they let the registration fee slide in order to pay the tickets, and the car gets impounded, and they miss a parole appointment, and the man faces getting returned to prison. And he is pissed and arguing and drinking more and gets into a fight or impulsively decides to grab the money at a stop & rob, and something goes terribly wrong.

I once raised my speculation on the relation of car problems to murder at a conference of lawyers and investigators, and a ripple of recognition swept through the room. Throughout the weekend, people approached me to say that two or three of their recent clients had begun to fall apart after a similar problem a month or so before the killing at issue. The need for public transportation is not solely a matter of the utopian demands of bicyclists, or the commuting needs of suburban company employees. The emotional and material impact on poor and working people of the lack of public transportation is profound. Public transportation is literally a life-saving backup for the working poor. A bus riders union in Los Angeles has recently popularized the term "transit racism" to describe the current transit priorities of our cities.

Police Tales

Perhaps the most basic stories that I hear relate to encounters with the police. I am less antagonistic to police officers than I used to be, but I am more critical of police structures and police culture than ever. All cops and most mayors assert in every budget request that the *thin blue line* stands between civic culture and civic chaos. I urge people to simply recall some of the major stories of the recent months and years: sodomy with a nightstick in a New York station house; death by 41 bullets of an unarmed man extending his wallet and identification in New York; death by shooting of an unarmed off-duty security guard who refused to respond to a plainclothes cop who solicited crack; hundreds of cases "jeopardized" by false testimony and planted evidence in Los Angeles finally revealed by a vice cop caught stealing cocaine from an evidence locker; faked forensic work by a Pennsylvania police lab, also "jeopardizing" hundreds of cases; some New Orleans cops working as guards, couriers, and hit men for drug dealers; district attorneys hiding exculpatory evidence to help their prosecutions. Perhaps the saddest stories of all are those of families who have called the police for help in dealing with an intoxicated son, or a mentally disturbed relative, or an abused daughter, only to have the encounter end in the death by gunfire of the person for whom they sought help.

"Young man, you don't understand. It happens every day."

The repeated claim of the police, most criminology departments, and virtually all politicians, is that in relation to crime, criminals are the problem, and police, whatever their flaws—"and they have a hard job"—are part of the solution. The stories I have heard for 30 years suggest that in relation to street crime, criminals and the police are perhaps equally the problem. Any planning for livable, sustainable cities in the future must take on the radical revamping of the police function. Community control and direct accountability—which are not yet in place anywhere that I know of—are just the beginning.

Within every major police force there are good people who know how deeply internal and community reforms have to go. There are even cops who deplore the fact that police unions always defend their members, no matter how clear the horrors they have perpetrated. I have been told more than once by reformers within the police structures that they cannot get very far when the pressure from outside is so weak, and who feel that they can only be effective, as police and as reformers, if public pressure for a revamping of the police function dramatically intensifies.

Attitudes Toward "The Law"

Although there is general "middle class" acceptance of the legitimacy of the police and the centrality of the "rule of law," many perceive that the "rule of law" carries almost no legitimacy for either the rich and powerful or the poor and powerless. The families of men in prison are very alert to the stories of the unpenalized abuses of power by the well off, and they talk about it. This major civics discussion of the poor is bound to undermine for them the middle class "rule-of-law" platitudes presented in high school civics classes.

The general street consensus might have been most cogently articulated by Felix Mitchell, a Bay Area cocaine dealer who was written about in local magazines as having "corporatized" the cocaine trade a few years ago. Felix introduced formalized career ladders, performance goals and bonuses, use of modern accounting techniques, midmanagement incentives such as trips and condos, and training sessions. From his limousine—sporting Da Mob license plates—he distributed basketballs, sports tickets, and food in housing projects. What Felix failed to do was pay taxes. Like Al Capone, he finally went to prison for tax evasion. When asked why, with his sophisticated accountants, he hadn't covered himself, he replied that he had thought about it, and it was a matter of principle. "The U.S. Government is a gang, like my gang, only older and more organized. We don't pay tribute to other gangs. That's it."

When Felix was killed in prison, over a drug debt of less than a hundred dollars, there was a funeral procession in Oakland that stretched for miles. A few brave African-American ministers tried to publicly discuss the drain of natural leadership talent into the drug trade, and to suggest that a social reconciliation process had to extend in this country to natural leaders who were lost first to the drug trade in the absence of other viable social options, and then to outrageous terms in prison.

Those who spoke up were derided in all the newspapers, by politicians from both parties, by the police, and by the established black middle-class leadership. The discussion was silenced.

Prison Construction

The same urban and suburban social groups that consistently vote down needle exchange programs, money for rehabilitation programs, halfway houses, women's shelters, and of course anything placed close to their homes or businesses, instead vote to build more and more prisons in remote rural locations. This of course makes the maintenance of family and legal contact for prisoners all but impossible, which is occasionally admitted as one of the goals. It is also changing the nature of small towns and much of rural America.

In these towns across rural California or elsewhere, when I used to talk to waitresses or gas station attendants or school teachers, I would find that their families were rooted in the local mills, or railroads, or canneries. Today the families all seem to have at least one person working for the local prison, and they hope that this will lead to others getting in as well. These are the best, often the only, well-paying and secure jobs in town. This means a whole community develops an investment in social control, in becoming latter-day slave catchers. The Boy Scouts, the local Little League, the Lions Clubs, the breakfast shops—all the formal and informal institutions of midsized towns—come to be dominated by prison guards, and the concerns of the guard unions and the developments at the prison dominate social discussion. In this process, the expansion of social control mechanisms becomes accepted as perhaps the primary source of the new jobs of the deindustrialized working class.

Cities and the Role of Contrast

Another complex of stories that I hear repeatedly are those about how life in the cities (and the media representations of that life) taunts poor people by dangling the fantasies of lives that are beyond their reach. A significant number of the crimes that I investigate have been preceded within 2–3 days by an unusual outing: a visit to a fancy restaurant, a club or a party; occasionally a drive; sometimes a major family trip to a new mall or a boutique neighborhood. Often there is a birthday party or a wedding where gifts were central. Often the defendants are unable to articulate why, but they realize that they were somehow depressed and angry after the experience. Often the confusion about the sources of a mood swing is itself a contributor to the denouement that follows—petty arguments, mutual belittling in whatever social relationship is at play, spousal violence or violence toward children, drinking or drugs with a male friend. Serious violence.

These potential explosions that result from the contrasts of wealth and opportunity and the lack of them are yet another indication that deindustrialization is wiping out the Black working class and scattering the white, thereby undermining the institutions that created a community life that ameliorated some of the stress of facing the contrasts. Black churches, company sports teams, busy local restaurants, garages that were also informal community centers, parks filled with weekend picnics, the guys with retirement pensions who could volunteer in recreation programs and who had respect from local kids on the street, these former sites of hope are all casualties of deindustrialization, and the DOT COMS of the world are not replacing them. In most of the murders I investigate, the parents of defendants—whites as well as Latinos and African Americans—get around sooner or later to talking about the changes in community social structure they have experienced within just a few decades, and the isolation of their kids compared with the relatively dense social structure of their own youth.

A recent trial brought together many of these issues.[2] My client was a 19-year-old high-school dropout (Duane) who was being pressured by his drug-addicted and welfare-dependent single mother to help pay the rent since he was now old enough to share responsibility. One night he went for a walk with a couple friends. In the neighborhood in which he lived "to go for a walk" is understood to mean "do a robbery." "You wanna go for a walk? Better come strapped." (Just a couple of years ago, the term for a successful robbery was "to get paid." "How'd you do on that walk?" "I got paid." The term "to get paid" seems to have been dropped; perhaps "getting paid" is such a rare experience that it has lost even its metaphoric utility.)

When he was eighteen Duane had received some money from an insurance settlement for a toxic spill in the neighborhood, and for about 3

months he lived well. He bought his mother a turkey and his grandmother a robe and took his girlfriend to a hip hop concert. He had some
good clothes that caught the eye of the neighborhood; he bought a bunch
of CDs and gave some out; he rented a phat car on some false ID and
cruised for girls. But the money went quickly; he was now scraping by
again and the people who had shared his windfall were back on his case
as if he had never done anything for them. Duane had submitted appli-
cations to a Home Depot, a Jack-in-the Box, and a McDonald's within
about 3 months, and been turned down by them all because he had a
record for one hand-to-hand drug sale of a couple rocks of crack cocaine
(total worth, maybe $20.00). His girlfriend was angry because she had
loaned him $200 that she had gathered from relatives so he could buy a
car to look for work and take her to visit her cousin, who had recently
moved out of town when *her* mother had been evicted. The car had been
impounded when Duane was stopped for no tags and no license, and he
could not afford to pay the mounting storage fees to get it back.

On the "walk," Duane and his friends stopped some Asian youths
about their own age, who weren't wearing gang colors, and who were in
a nice neighborhood late at night, so they probably had some money on
them—to party, for 40s, or for drugs. They looked like easy marks: no
likely resistance, quick and no trouble, no violence. Duane had done this
only two times before and was nervous, not sure at all he wanted to go
through with it, but one of his younger friends was just out of Juvie, and
was pushing the robbery because he wanted the experience as much as he
wanted the money. They walked around the block a couple of times to
make sure things were clear and then there was a quick confrontation
and a demand for money.

It went wrong, of course. Duane's friend and one of the Asian youths
got into a heated exchange. A second kid moved toward Duane and
stuck his hand inside his jacket. Nerves fired, someone pulled a trigger,
and the young man that Duane had initially confronted fell dead. Truthfully, I am not yet sure exactly what happened in those few seconds
because none of the stories of the six people involved are exactly the
same, and the physical evidence seemed to me to be inconclusive. Duane
was arrested a few days later and was soon facing a death penalty trial.

The crime was of course horrible, and a life was lost and wasted. The
young victim left a large, vibrant, and loving family who came to court

regularly. They contrasted sharply with Duane, who was sullen and withdrawn, and his own relatives, who were confused, angry, and generally absent from the trial. The defense team, the bailiff, and the judge all thought the defense had raised serious doubts about who the shooter was and what had really happened, but Duane was found guilty within an hour and a half. The jury clearly did not like him, dismissed the defense, and was in a vengeful mood.

There are two stages in capital cases in California and in most states. In the first, a jury assesses an incident to make a determination of guilt. In the second, or penalty phase, the jury is supposed to assess a life as a whole to make a determination of the penalty: execution or life without the possibility of parole (LWOPP). In the second phase we, the defense team, can put before the jury almost anything we think might help them understand the life of our defendant. Our task was to help the jury understand Duane as the product of his circumstances and his community, if they were to understand him at all.

The lawyers initially wanted to rely on a psychologist and a cultural geographer as the core of the defense. They didn't think that Duane's family would be appealing witnesses; they had never used lay community people in a trial before and they wanted to rely on sympathetic experts to help the jurors understand the defendant in his social context. We came to agree that primarily jurors needed to hear stories, and that experts are only as strong as the stories upon which their assessments are based. Without the raw stories, information from experts could be considered biased. If the stories are strong and direct, experts then are not positioning themselves between jurors and defendants, or providing information. Rather, they come to assist jurors in assessing the stories they have both heard.

Ultimately, I was the first defense witness. I presented a long slide show, and the attorneys took me through a long story (which we had scripted) about the process of my own investigation of Duane and his community. The slides were about life in the neighborhood, in which I tried to depict the isolation and bleakness that this kid walked through every day of his life. But the deeper story was about the process itself of taking in and learning about the community, and the difficulty of getting beyond the superficial, as compelling as it was. We wanted to engage the jurors from the beginning in the process of exploration of this life.

Instead of the professional cultural geographer, we put family members on the stand, not so they could shine and be sympathetic, but so that the jurors could see them in all their dysfunctions and confusion. They did not disappoint.

I had spent a lot of time building relationships in the community and so we were able to bring in old men to talk of the changes in their block and neighborhood, ministers to talk about the changes in the church, teachers to talk about their own decisions to teach in these schools and the challenges they faced there, nurses to talk about nights in the local hospitals. We drew in the people who had organized campaigns against toxic waste and for bus routes to their isolated neighborhood. We drew on the "needs assessments" prepared for health centers and community-center grant submissions to guide questioning about neighborhood conditions and history.

More, we brought convicted ex-gang members to talk about why and how they began to use and sell drugs and what gangs really were in their lives. We used a minister who talked about his three convictions for assault, a retired utilities worker who volunteers in a truancy-prevention program, a retired lounge owner who leads antiviolence theater groups among young men at a community center, a woman who left the stand in tears as she described how she got burned out working in a community center and failed to reach the kids she loved.

In short, we let people tell their own stories, as well as talk about what they knew of our client (which was often very little), and we let the jury make the connections among the stories, the photos, the family, the community, and the defendant. The witnesses moved easily and effectively from their own stories to those of their institutions and communities; their anecdotes were compelling and their analyses sharp. Lawyers and a host of other middle-class professionals, academics included, generally underestimate the capacities of poor and working class people to think and speak for themselves. The glittering lights of the faculty club often blind rather than illuminate.

Many of the people—men and women—who work in community institutions, many of our witnesses, were themselves former addicts, had histories of inflicting or receiving severe violence, and had done time in prison. The possibility of change and transformation was implicit in their stories. Yet if the severe "sentencing reforms" recently and widely enacted at both federal and state levels had been in place even 10 years ago, many of them would still be in prison, a couple of them for life. They would be rotting away, rather than working actively to transform their communities.

The people we presented were those who most felt the impact of the violence in their community. They had raised, worked and lived with, taught, and loved kids who had been shot, and they acknowledged their grief and anger at the losses. Yet they were also people who felt the loss of another kid—our defendant—to the streets, to violence, and then to prison and/or execution, as deeply as they felt the loss of those who bled to death in the streets. Both these losses were seen and discussed as part of the same dynamic—the process of life in the so-called inner cities of America.

The jury, clearly angry and inclined to a death verdict at the beginning of the penalty phase of the trial, needed less than an hour's deliberation to decide that there was no need to kill this young man, with whose life they now empathized. It was a stunningly swift verdict for life, a verdict unheard of in this county.

This 19-year-old defendant also heard the stories and began to think of himself and his life differently. His face, like those of the jurors, turned softer and more attentive over the course of a few days. He too began to place his life in a social context, perhaps also for the first time. It is sad that a murder and a trial had to occur before this process of reflection could take place. Now he is off for life to a place where the very notion of rehabilitation has been dropped, to the most racially segregated institution in the world, where racial identity and a commitment to racial violence are reinforced daily.

It is encouraging that jurors can often hear the stories of defendants' lives in ways that can lead them to think critically about what conditions behavior, and that this understanding in turn may help them to both temper their anger and to think more expansively about their communities. Story may also help us save and shape these communities, but only if we value but do not overstate its significance. At their core, these are stories about power relationships, and speaking truth to power is generally a fruitless exercise unless it is linked to building counterpower. Without this connection, stories may become mere journalistic titillation or academic voyeurism.

The same people who came forward and helped us situate this one life within a broader collective context are out there, in every city in the country, struggling on a daily basis to make our cities livable and our society more sane and humane. Some of them are trying to find ways to broaden their connections and unite their efforts for change, to overcome old antagonisms, to find places to talk and unite with those who bring other skills and insights to the work of social transformation, to hear other stories on a basis of mutual respect.

Story at its best creates a sense of commonality; commonality generates a sense of shared history; and a sense of shared history, even if it is a contested history, creates the possibility of community. We need to create spaces and ways to hear, share, draw lessons from, and *act on* these stories, *before* we are called for jury duty or hauled before a court.

Notes

Chapter 1

1. I thank John Forester and Jim Throgmorton for their generous suggestions for improving this essay and Patsy Healey for her encouraging words.

2. Although I know that "citizens," "residents," and "stakeholders" are the usual terms planners employ to specify people in a city, I follow Michel de Certeau (1984) and Guido Martinotti (1999) and employ the word "users." In doing so, I mean to include temporary city occupants, such as tourists, conventioneers, and temporary workers, as well as homeless people and illegal aliens. Certainly, they all have something at stake in city plans, but I do not often hear them included when the term "stakeholders" is used.

In an effort to follow the disciplinary practice that I see in major planning journals, I use the word "planner," but I recognize that individuals with many different kinds of knowledge and with many different kinds of jobs are called upon or offer their efforts to city planning projects. Not all of these individuals went to planning schools or call themselves planners. By "planner" I mean anyone engaged in urban planning broadly conceived.

3. What Plumwood calls consequential remoteness, economists and social choice theorists have debated in terms of externalities: negative externalities being the unpriced costs borne by persons socially, spatially, and temporally distant from the principal actors; and positive externalities being the unpriced benefits enjoyed by those central principal actors. In her most recent book, *Environmental Culture: The Ecological Crisis of Reason* (2002), she returns to her argument that small autarchic communities are not the answer to consequential remoteness and links this term to the discourse on externalities. "In place of the extreme autarchic solution of eliminating exchange in order to eliminate remoteness, there is thus the option of restructuring exchange so that it maintains equivalent levels of remoteness. . . . In place of centrist forms of exchange that export negative externalities from the centre and import positive ones from the periphery, fair exchange aims for a balanced distribution of broadly conceived costs and benefits so that they fall equally on both sides" (Plumwood 2002, p. 79).

4. In the 1970s Portelli collected oral memories of the shooting of Luigi Trastulli during a 1949 labor and Communist Party–organized anti-NATO protest. What

he found is a meaningful pattern of errors in memory about the date and the nature of the protest. Portelli surmises that the workers who remembered the Trastulli incident narrated a tale in which Trastulli was killed in a later protest over lost jobs because, by the 1970s, resistance to NATO was no longer a central tenet of the Italian Communist Party. Yet, Trastulli remained a martyr in the collective memory of the labor community which was increasingly threatened by deindustrialization.

5. If "we" listen to National Public Radio, we might have heard LeAlan Jones and Lloyd Newman's radio essays earlier in the 1990s.

6. I would like to say "train," but there are no passenger trains within 50 miles of where I live.

Chapter 2

1. I thank Robert Beauregard, Leonie Sandercock, Gerald Frug, Seymour Mandelbaum, Barbara Eckstein, and participants in the rhetoric seminar of the Project on Rhetoric of Inquiry for offering their comments on earlier versions of this essay. Thanks too go to Sara Walz for diligently responding to my persistent requests for help in tracking down details.

2. For good introductions to that discourse, see Campbell (1996), Myerson and Rydin (1996), Peterson (1997), Torgerson (1999), and Davison (2001). In the United States, terms such as "smart growth" and "livable communities" are often used in place of "sustainability," largely because many Americans (especially property-rights advocates) associate sustainability with a left-wing environmentalism.

3. I would like to thank Barbara Eckstein for pointing out the relevance of Jones and Newman's tale to my argument.

4. As Brian Ladd puts it in *The Ghosts of Berlin* (1997, p. 1), "Memories often cleave to the physical settings of events. That is why buildings and places have so many stories to tell. They give form to a city's history and identity."

5. See DeLeon (1992) for an insightful story about the effort of San Francisco's "growth regime" to produce "sustained growth." Sayre, Oklahoma, exemplifies another version, one that reveals what sustained growth efforts can lead to if they are not tempered by other considerations. Shattered by the collapse of the oil and gas business in the 1980s, and just barely surviving on federal crop support payments, this town of 4,000 people located 120 miles west of Oklahoma City is trying to build a sustainable growth economy based on a privately operated prison, a meat-processing plant, and (they hope) a Wal-Mart distribution center. The inmates of the prison are shipped to Sayre from Wisconsin, which has a space shortage in its own prisons (Kilborn, 2001).

6. One example involves New York City and Sierra Blanca, Texas, a town of 600 people about 90 miles southeast of El Paso. After Congress banned ocean dumping of New York's sludge in 1992, the city contracted with private firms to dispose of "biosolids" left after treatment of its sewage. One of the firms has

begun shipping up to 250 tons of sludge 2,065 miles by rail to Sierra Blanca every day. Some local residents complain of health risks and environmental injustice, whereas others reply that all environmental standards are met and the disposal site provides needed jobs and income (Yardley, 2001).

Another example involves energy transmission. The proposed "millenium pipeline" would carry natural gas 425 miles from Lake Erie to New York City, where it would be used to generate electricity. At Lake Erie, the pipeline would connect with a proposed 180-mile segment in Ontario, which in turn would connect with a transcontinental network transporting gas from western Canada. The proposed pipeline has stimulated broad-based opposition. One of the opponents said, "It's not just a Croton issue [where she lives], it's not just a Westchester County issue. It's a regional issue. It's a New York State issue. It's a New York City watershed issue, a Great Lakes issue" (Archibold, 2001, p. A12).

7. Jonathan Harr's (1996) story about Woburn, Massachusetts, exemplifies the ways in which remote adverse effects of tentacular radiations can become submerged in a people's environmental unconscious but then emerge again in ways that reawaken them to our interdependence with other places and times.

8. By claiming that we in "the North" should create environmental space for development in "the South," they are making a strong moral claim. In *Edge City*, Joel Garreau (1991) writes, "Put aside the holes in the ozone and craziness in the Middle East for one moment, for the sake of argument. For better or for worse, there are no geological reasons that plenty of oil should not be available throughout the twenty-first century. . . . If *we* choose to accept the political and environmental and military and economic consequences, *we* can find enough oil on this planet" (1991, pp. 125–126; emphasis added). Who is the "we" he refers to? What makes us Americans think that we have a right to oil from other parts of the world? In what sense do "we" Americans have any right to "choose to accept" political and environmental and military and economic consequences borne by other people in other parts of the world?

9. The literature about the New Urbanism is large and growing. For good introductions, see Kunstler (1996), Duany et al. (2000), and the Congress for the New Urbanism (2000).

10. Such considerations of physical design lead many an urban planner to want American cities to be more like European cities (e.g., Beatley, 2000).

11. This fear also applies to transnational institutions, such as the G-8, the International Monetary Fund, the World Bank, and the World Trade Organization, who promulgate a globalization of the economy that benefits the "free trade" interests of transnational corporations. The many thousands of people who demonstrated in the streets of Seattle, Washington (1999), Genoa, Italy (2001), and elsewhere clearly believe that these organizations need to be restructured or completely replaced with ones that are more democratic. Their objective is not so much to challenge economic globalization as it is to expand the range of voices that shape how globalization occurs.

12. Torgerson (1999) also argues that sustainability is something that should be created discursively through democratic political practice. But he also urges

advocates of sustainability to recognize the *comic* dimension of their efforts. "[I]f ecology indicates the interconnection, complexity, and significant unpredictability of nature," he writes, "we can also hold up the mirror of ecology to reveal the corresponding character of human affairs. The intricate, continually surprising maze of politics and history compels us to think always with a question mark. We must act with conviction, but with the understanding that we might be (somehow must be) wrong" (1999, p. 103).

Chapter 3

1. This chapter has benefited significantly from the advice of Barbara Eckstein, Judith Garber, Meg Holden, Robert Lake, Daphne Spain, and James Throgmorton.

Chapter 4

1. My thanks to experts, old neighbors, family members, friends, and accommodating strangers who took time to talk to me about South Shore, especially Carol and Emil, Darryl and Tonia, Rhonda McMillon, Noreen and Gilbert Cornfield, Gayle Pemberton, Richard Taub, Bob Keeley, Commander James Polk of the Chicago Police Department, Father William Sheridan of St. Philip Neri, Ron Grzywinski and Milton Davis of the South Shore Bank, and Tiffany Charles of the Chicago Historical Society. My parents, Salvatore and Pilar Rotella, patiently answered my questions and provided hairsplitting commentary on a draft of this essay. Thanks also to my brothers, Sebastian and Sal, for their comments on a draft, and for preventing that kid in the horses' heads pajama top from getting my baseball glove. I extend my appreciation, finally, to the Passmans, Thigpens, Trainers, and Earps: good neighbors in the truest, best sense of the words.

2. These measures of population and income, dating from the late 1990s, come courtesy of two gracious experts on South Shore, Sherry Herman and Chris Berry of ShoreBank's Neighborhood Marketing Initiative.

3. For discussion of the city of feeling and the city of fact, see Rotella (1998), especially pp. 1–16. For a preliminary consideration of the relationship between neighborhood as artifact and as quality, see pp. 123–124.

4. Farrell (1993 [1932, 1934, 1935]). For scenes that place Lonigan at 71st and Jeffery, see especially pp. 687 and 797. For an extended discussion of Chicago realism as a literary tradition, see Rotella (1998, pp. 19–115).

5. See Pemberton (1992, pp. 179–181 and p. 193).

6. See Molotch (1972, p. 229).

7. The estimate of 12,000 rehabilitated apartments appears in McRoberts and Pallasch (1999, pp. 1 and 12–13).

8. Again, see Molotch (1972, pp. 228–229).

9. Hattie Wilburn's letter, handwritten in "careful cursive on sheets of yellow legal paper," is quoted in McRoberts and Pallasch (1999, p. 1).

10. See Suarez (1999, pp. 21–23).

11. See Suarez (1999, p. 44).

12. See Molotch (1972); Taub, et al (1984); Taub (1988). Taub, Taylor, and Dunham (1984) outline "the classic model," pp. 38–75.

13. Molotch (1972, p. 45).

14. I saw "The Old Neighborhood" on Broadway in December 1997. Quoted lines are as they appear in the published script (Mamet, 1999). Mamet has also contributed to the "71st and Jeffery!" subgenre of South Shore stories. His essay "71st and Jeffrey [sic]" (1994, no page numbers) begins with an idyllic account of his life as a boy in Jewish South Shore, playing all day in the streets, watching cartoons at the movie house all day for a quarter, scarfing chocolate phosphates and Francheezies. It ends, years later after Mamet has moved to the North Side, with a violent crime: he is robbed at knifepoint on the North Side while driving a Yellow Cab. The cop who takes the robbery report turns out to be the new owner of the old Mamet house in South Shore. They agree that the robber is long gone, then Mamet drives off in his cab—"and that," he concludes, "was my last connection with the old neighborhood."

15. See Cronin (1968, p. 5).

16. See Sharoff (2000, p. 27).

Chapter 5

1. For more on the East St. Louis Action Research Project, see www.eslarp.uiuc .edu.

Chapter 6

1. An old friend and someone whose practice has always inspired me.

Chapter 7

1. This essay derives from an oral interview with Michael Berkshire conducted by James A. Throgmorton in Throgmorton's home in Iowa City on April 21, 2001. With Berkshire's approval, the editors subsequently transcribed and revised the interview. Berkshire read and approved the final version.

2. The ECICOG solid waste region includes the counties of Benton, Iowa, Johnson, Jones, Linn, and Tama.

3. For details about Bluestem, see their web site at http://www.bluestem.org.

4. While still an employee of ECICOG, Liz Christiansen was hired contractually to manage the formation of Bluestem. She was then hired as the Planning and Education Coordinator for Bluestem.

Chapter 8

1. For a wonderful guide to these systems, see Brail and Klosterman (2001).

2. On the temporal structure of plans, see Hopkins (2001).

3. These brief notes on the repertoire as resource were first prepared as an oral prologue to a critical reading of the case studies commissioned by the organizers of the Iowa symposium. I had the luxury of reading the cases in advance and was (or so it seemed initially) the agent for my colleagues, who would only hear the cases without a written text in hand. I was a tutor as well since I had the advantage of reading rather than "merely" listening. My confidence in that putative advantage quickly vanished. Listening to the narrators, I found a passion and a humor in their oral accounts that was muted in the written text and in the protocols of reading. That was the first lesson of a symposium on narratives and (it surely followed) on narrators.

Chapter 9

1. Most new residential subdivisions have housing on both sides of the street. They are double loaded. Single-loaded streets have housing on only one side.

2. Live-work units combine places of residence and work; such buildings typically contain a business (e.g., a cafe) on the first floor of a primary residence.

3. For further details about the Peninsula project and its current status, see http://www.thepeninsulaneighborhood.com.

Chapter 10

1. Eckstein specifically refers to the oral storytellers and historians of Inuit society.

Chapter 11, Editors' Introduction

1. The editors thank Doris Witt for suggesting some of these references to prison scholarship. Critical Resistance can be found at http:/www.criticalresistance.org.

Chapter 11

1. "40s" is as common a term as "6 pack," "spritzer," or "martini glass." The term "40" is clear to anyone who drinks or sells beer or malt liquor.

2. Some names and identifiers have been changed.

References

ACSP Update. (2000). Nov/Dec. Milwaukee, Wis.: Association of Collegiate Schools of Planning.

Archibold, Randal C. (2001). "Gas Pipeline Faces Opposition: Not Under My Backyard." *New York Times* (August 7), p. A 12.

Athanasiou, Tom. (1996). *Divided Planet: The Ecology of Rich and Poor.* Boston: Little, Brown.

Baer, William C. (1997). "General Plan Evaluation Criteria: An Approach to Making Better Plans." *Journal of the American Planning Association,* 63, 329–344.

Bakhtin, M. M. (1981). Forms of Time and of the Chronotope in the Novel. In Michael Holquist (ed.), *The Dialogic Imagination: Four Essays* (Caryl Emerson and Michael Holquist, trans., pp. 84–258). Austin: University of Texas Press.

Baldwin, James. (Rpt. 1965, from 1957). Sonny's Blues. In *Going to Meet the Man.* New York: Vintage Press.

Barthes, Roland. (1973). *Le Plaisir du Text.* Paris: Seuil.

Baum, Howell, S. (1997). *The Organization of Hope.* Albany: State University of New York Press.

———. (1999). "Forgetting to Plan." *Journal of Planning Education and Research,* 19, 2–14.

Beatley, Timothy. (2000). *Green Urbanism: Learning from European Cities.* Washington, D.C.: Island Press.

Beatley, Timothy, and Kristy Manning. (1997). *The Ecology of Place: Planning for Environment, Economy, and Community.* Washington, D.C.: Island Press.

Beauregard, Robert A. (1991). "Without a Net: Modernist Planning and the Postmodern Abyss." *Journal of Planning Education and Research,* 10, 189–194.

———. (1995). If Only the City Could Speak: The Politics of Representation. In H. Liggett and D. C. Perry (eds.), *Spatial Practices* (pp. 59–80). Thousand Oaks, Calif.: Sage.

———. (1999). "Julkinen kaupunki" (The Public City). *Janus,* 7, 214–223.

Beauregard, Robert A., and Anna Bounds. (2000). Urban Citizenship. In E. Isin (ed.), *Democracy, Citizenship and the Global City* (pp. 243–256). London: Routledge.

Bellah, Robert N., R. Madsen, W. M. Sullivan, A. Swidler, and S. M. Tipton. (1991). *The Good Society*. New York: Vintage.

Bender, Thomas. (2001). "The New Metropolitanism and the Pluralized Public." *Harvard Design Magazine*, 13, 70–77.

Berger, John. (1974). *The Look of Things*. New York: Viking.

Berman, Marshall. (1982). *All That Is Solid Melts into Air*. New York: Penguin Books.

Bernick, M., and R. Cervero. (1996). *Transit Villages in the 21st Century*. New York: McGraw-Hill.

Bohman, James. (1996). *Public Deliberation: Pluralism, Complexity, and Democracy*. Cambridge, Mass.: MIT Press.

Borges, Jorge Luis. (1964). The Garden of Forking Paths (Donald A. Yates, trans.). In *Labyrinths: Selected Stories and Other Writings*. New York: New Directions.

———. (1971). *The Aleph and Other Stories: 1933–1969* (Norman Thomas di Giovanni in collaboration with the author, trans.). New York: Bantam Books.

Bourdieu, Pierre. (Rpt. 1993, from 1980). The Production of Belief: Contribution to an Economy of Symbolic Goods (French, 1977; Richard Nice, trans., 1980). In Randal Johnson (ed.), *The Field of Cultural Production: Essays on Art and Literature* (pp. 74–111). New York: Columbia University Press.

———. (Rpt. 1993, from 1983). The Field of Cultural Production, or: The Economic World Reversed (Richard Nice, trans.). In Randal Johnson (ed.), *The Field of Cultural Production: Essays on Art and Literature* (pp. 29–73). New York: Columbia University Press. From *Poetics* (Amsterdam) 12/4–5 (1983), pp. 311–356.

———. (2000). *Pascalian Meditations* (Richard Nice, trans.). Stanford, Calif.: Stanford University Press.

Boyer, M. Christine. (1994). *The City of Collective Memory: Its Historical Imagery and Architectural Entertainments*. Cambridge, Mass.: MIT Press.

Brail, Richard K., and Richard E. Klosterman, eds. (2001). *Planning Support Systems: Integrating Geographic Information Systems, Models, and Visualization Tools*. Redlands, Calif.: ESRI Press.

Brooks, Peter. (2001). "Stories Abounding." *Chronicle of Higher Education* (March 23) 47, B11.

Buell, Lawrence. (2001). *Writing for an Endangered World: Literature, Culture, and Environment in the U.S. and Beyond*. Cambridge, Mass.: Belknap/Harvard University Press.

Bullard, Robert. (1994). *Dumping in Dixie: Race, Class and Environmental Quality*, 2nd ed. Boulder, Colo.: Westview.

Caldeira, Teresa Pires do Rio. (1996). "Fortified Enclaves: The New Urban Segregation." *Public Culture*, 8, 303–328.

Calhoun, Craig. (1993). "Civil Society and the Public Sphere." *Public Culture*, 5, 267–280.

Calthorpe Associates. (1990). *Transit-oriented Development Design Guidelines*. Calthorpe Associates, Sacramento County Planning and Community Development Dept.

Calthorpe, Peter. (1993). *The Next American Metropolis: Ecology and Urban Form*. New York: Princeton Architectural Press.

Calthorpe, Peter, and William Fulton. (2001). *The Regional City: Planning for the End of Sprawl*. Washington, D.C.: Island Press.

Campbell, Scott. (1996). "Green Cities, Growing Cities, Just Cities? Urban Planning and the Contradictions of Sustainable Development." *Journal of the American Planning Association*, 62, 296–312.

Carson, Rachel. (1962). *Silent Spring*. Boston: Houghton Mifflin.

Castells, Manuel. (1997). *The Power of Identity*. Malden, Mass.: Blackwell.

Certeau, Michel de. (1984). *The Practice of Everyday Life* (Steven Rendall, trans.). Berkeley: University of California Press.

Chafe, William H. (1986). *The Unfinished Journey: America Since World War II*. New York: Oxford University Press.

Charter of European Cities and Towns towards Sustainability (the Aalborg Charter). (1994). Approved by participants at the European Conference on Sustainable Cities and Towns, Aalborg, Denmark.

Congress for the New Urbanism. (2000). *Charter of the New Urbanism*. New York: McGraw-Hill.

Cronin, Pat Somers. (1968). "The Agony of South Shore." *Daily News Panorama* (February 4), p. 5.

Cronon, William, ed. (1996). *Uncommon Ground: Rethinking the Human Place in Nature*. New York: W. W. Norton.

Dagger, Richard. (1997). *Civic Virtues*. New York: Oxford University Press.

Dale, Norman. (1999). Cross-Cultural Community Based Planning. Negotiating the Future of Haida Gwaii. In L. Susskind, S. McKearnan, and J. Thomas-Larmer (eds.), *The Consensus Building Handbook* (pp. 923–950). Thousand Oaks, Calif.: Sage.

Davis, Angela Y. (1971). *If They Come in the Morning*. New York: Third Press.

———. (2001). "Prison as a Border: A Conversation on Gender, Globalization, and Punishment." *Signs: Journal of Women in Culture and Society*, 26, 12–35.

Davison, Aidan. (2001). *Technology and the Contested Meanings of Sustainability*. Albany, State University of New York Press.

DeLeon, Richard Edward. (1992). *Left Coast City: Progressive Politics in San Francisco, 1975–1991*. Lawrence: University Press of Kansas.

Drake, St. Clair, and Horace R. Cayton. (1945). *The Black Metropolis: A Study of Negro Life in a Northern City*. New York: Harcourt, Brace.

Duany, Andres. (1998). "Our Urbanism." *Architecture*, 87, 37–40.

Duany, Andres, Elizabeth Plater-Zyberk, and Jeff Speck. (2000). *Suburban Nation: The Rise of Sprawl and the Decline of the American Dream*. New York: North Point Press.

Dubrow, Gail. (1998). Feminist and Multicultural Perspectives on Preservation Planning. In L. Sandercock (ed.), *Making the Invisible Visible. A Multicultural Planning History* (pp. 57–97). Berkeley: University of California Press.

Dutton, John A. (2000). *New American Urbanism: Re-forming the Suburban Metropolis*. Milan, Italy: Skira.

Ehrenhalt, Alan. (1995). *Lost City: The Forgotten Virtues of Community in America*. New York: Basic Books.

Ellison, Ralph. (1952). *Invisible Man*. New York: Random House.

Energy Information Administration. (1997). Petroleum 1996: *Issues and Trends*. DOE/EIA-0615. Washington, D.C.: U.S. Energy Information Administration.

Energy Information Administration. (1998). Deliverability on the Interstate Natural Gas Pipeline System. DOE/EIA-0618(98). Washington, D.C.: U.S. Energy Information Administration.

Farrell, James T. (Rpt. 1993, from 1932, 1934, 1935). *Studs Lonigan: A Trilogy Comprising Young Lonigan, The Young Manhood of Studs Lonigan, and Judgment Day*. Urbana: University of Illinois Press.

Faulkner, William. (Rpt. 1956, from 1929). *The Sound and the Fury*. New York: Modern Library.

Ferraro, Giovanni. (1994). "*De te fabula narratur*. Exercices [sic] in Reading Plans." *Planning Theory*, 10/11, 205–236.

Finnegan, Ruth. (1998). *Tales of the City: A Study of Narrative and Urban Life*. New York: Cambridge University Press.

Flyvbjerg, Bent. (1998). *Rationality and Power: Democracy in Practice*. Chicago: University of Chicago Press.

———. (2001). *Making Social Science Matter: Why Social Inquiry Fails and How It Can Succeed Again*. Cambridge: Cambridge University Press.

Forester, John. (1989). *Planning in the Face of Power*. Los Angeles: University of California Press.

———. (1999). *The Deliberative Practitioner: Encouraging Participatory Planning Processes*. Cambridge, Mass.: MIT Press.

———. (2000). Multicultural Planning in Deed: Lessons from the Mediation Practice of Shirley Solomon and Larry Sherman. In M. Burayidi (ed.), *Urban Planning in a Multicultural Society*. Westport, Conn.: Praeger.

Foucault Michel. (1980). Questions on Geography. In *Power/Knowledge* (C. Gordon, trans., pp. 63–77). New York: Pantheon.

————. (1986). "Of Other Spaces" (Jay Miskowiec, trans.). *Diacritics*, 16, 22–27.

Franklin, H. Bruce. (1998). *Prison Writing in Twentieth-Century America*. New York: Penguin Books.

Fraser, Nancy. (1995). "From Redistribution to Recognition: Dilemmas of Justice in a 'Post-Socialist' Age." *New Left Review*, 212, 68–93.

Frug, Gerald E. (1999). *City Making: Building Communities without Building Walls*. Princeton, N.J.: Princeton University Press.

Gale, Dennis E. (1996). *Understanding Urban Unrest*. Thousand Oaks, Calif.: Sage.

Garber, Judith A. (2000). The City as an Heroic Public Sphere. In E. Isin (ed.), *Democracy, Citizenship and the Global City* (pp. 257–274). London: Routledge.

Garreau, Joel. (1991). *Edge City: Life on the New Frontier*. New York: Anchor Books.

Gennette, Gerard. (1971). Time and Narrative in *A la rescherche du temps perdu* (Paul de Man, trans.). In J. Hillis Miller (ed.), *Aspects of Narrative*. New York: Columbia University Press.

Goheen, Peter G. (1998). "Public Space and the Geography of the Modern City." *Progress in Human Geography*, 22, 479–496.

Gold, Michael. (1935). *Jews without Money*. New York: International Publishers.

Habermas, Jurgen. (1996a). *Between Facts and Norms* (William Rehg, trans.). Cambridge, Mass.: MIT Press.

————. (1996b). Three Normative Models of Democracy. In S. Benhabib (ed.), *Democracy and Difference* (pp. 21–30). Princeton, N.J.: Princeton University Press.

Harr, Jonathan. (1996). *A Civil Action*. New York: Random House.

Harrill, Rich. (1999). "Political Ecology and Planning Theory." *Journal of Planning Education and Research*, 19, 67–75.

Harvey, David. (1996). *Justice, Nature and the Geography of Difference*. Oxford: Blackwell.

————. (2000). *Spaces of Hope*. Berkeley: University of California Press.

Haughton, Graham. (1999). "Environmental Justice and the Sustainable City." *Journal of Planning Education and Research*, 18, 233–243.

Hawken, Paul, Amory Lovins, and L. Hunter Lovins. (1999). *Natural Capitalism: Creating the Next Industrial Revolution*. Boston: Little, Brown.

Hayden, Dolores. (1996). *The Power of Place*. Cambridge, Mass.: MIT Press.

Healey, Patsy. (1997). *Collaborative Planning*. London: Macmillan.

Hirsch, Arnold. (1983). *Making the Second Ghetto: Race and Housing in Chicago, 1940–1960*. New York: Cambridge University Press.

Holston, James, ed. (1999). *Cities and Citizenship*. Durham, N.C.: Duke University Press.

hooks, bell, and Cornell West. (1991). *Breaking Bread: Insurgent Black Intellectual Life*. Cambridge, Mass.: South End Press.

Hopkins, Lewis D. (2001). *Urban Development: The Logic of Making Plans*. Washington, D.C.: Island Press.

Houghton, J. T., Y. Ding, D. J. Griggs, M. Noguer, P. J. van der Linden, and D. Xiaosu. (2001). *Climate Change 2001: The Scientific Basis. Contribution of Working Group I to the Third Assessment Report of the Intergovernmental Panel on Climate Change*. Cambridge: Cambridge University Press.

Innes, J. (1995). "Planning Theory's Emerging Paradigm: Communicative Action and Interactive Practice." *Journal of Planning Education and Research*, 14, 183–190.

Iowa City Press-Citizen. (2001). "ICAD plan could take us forward." *Iowa City Press-Citizen* (June 11), p. 9A.

Isin, Egin. (1999). "Introduction: Cities and Citizenship in a Global Age." *Citizenship Studies*, 3, 165–171.

Isserman, Andrew. (1984). "Projection, Forecast, and Plan: On the Future of Population Forecasting." *Journal of the American Planning Association*, 50, 208–221.

Jacobs, Jane. (1961). *The Death and Life of Great American Cities*. New York: Modern Library.

———. (2000). *The Nature of Economies*. New York: Modern Library.

Jones, LeAlan, and Lloyd Newman (with David Isay). (1997). *Our America: Life and Death on the South Side of Chicago*. New York: Scribner.

Katz, P. (1994). *The New Urbanism: Toward an Architecture of Community*. New York: McGraw-Hill.

Katznelson, Ira. (1996). *Liberalism's Crooked Circle*. Princeton, N.J.: Princeton University Press.

Kelman, Steven. (1988). Why Public Ideas Matter. In R. B. Reich (ed.), *The Power of Public Ideas* (pp. 31–53). Cambridge, Mass.: Ballinger.

Kent, T. J. (1964). *The Urban General Plan*. San Francisco: Chandler.

Kilborn, Peter J. (2001). "Rural Towns Turn to Prisons to Revive Their Economies." *New York Times* (August 1), p. A 1.

King, M. (1981). *Chain of Change*. Boston: South End Press.

Krumholz, Norman, and John Forester. (1990). *Making Equity Planning Work: Leadership in the Public Sector*. Philadelphia, Pa.: Temple University Press.

Kunstler, James Howard. (1996). *Home from Nowhere*. New York: Simon and Schuster.

Ladd, Brian. (1997). *The Ghosts of Berlin: Confronting German History in the Urban Landscape*. Chicago: University of Chicago Press.

Lake, Robert W. (2000). Contradictions at the Local Scale. In N. Low, B. Gleeson, I. Elander, and R. Lidskog (eds.), *Consuming Cities* (pp. 70–90). London: Routledge.

Lamont, William, Jr. (1990). "Turning a Plan into History." *Journal of the American Planning Association* 56 (3): 356–358.

Lefebvre, Henri. (1991). *The Production of Space* (Donald Nicholson-Smith, trans.). Oxford UK and Cambridge, Mass.: Blackwell.

———. (1996). *Writings on Cities* (Eleonore Kofman and Elizabeth Lebas, trans. and eds.). London: Blackwell.

Lindblom, Charles E. (1977). *Politics and Markets*. New York: Basic Books.

Lofland, Lyn H. (1998). *The Public Realm: Exploring the City's Quintessential Social Territory*. New York: Aldine.

Loukaitou-Sideris, A. (2000). The Byzantine-Latino Quarter, *DISP* [Dokumente und Informationen zur Schweizerischen Orts-, Regional- und Landesplanung: Documents and Information on Swiss Local, Regional and State Planning (January), 1.

Mamet, David. (1994). 71st and Jeffrey (sic). In Tom Maday and Sam Landers (eds.), *Great Chicago Stories: Portraits and Stories* (no page numbers). Chicago: TwoPress Publishing.

———. (1999). *The Old Neighborhood*. New York: Random House.

Mandelbaum, Seymour. (1984). "Temporal Conventions in Planning Discourse." *Environment and Planning B: Planning and Design*, 11, 5–13.

———. (1990). "Reading Plans." *Journal of the American Planning Association*, 56, 350–356.

———. (1991). "Telling Stories." *Journal of Planning Education and Research*, 10, 209–214.

Marable, Manning. (1991). *Race, Reform, and Rebellion: The Second Reconstruction in Black America, 1945–1990*, 2nd ed. Jackson: University Press of Mississippi.

———. (2000). *How Capitalism Underdeveloped Black America: Problems in Race, Political Economy, and Society*, 2nd ed. Cambridge, Mass.: South End Press.

Marris, Peter. (1997). *Witnesses, Engineers, and Storytellers*. College Park, Md.: Urban Studies and Planning Program, University of Maryland.

Martino, Ernesto de. (1953). "Note di viaggio" (Alessandro Portelli, excerpt trans.). *Nuovi Argomenti*, 2, 47–79.

Martinotti, Guido. (1999). A City for Whom? Transients and Public Life in the Second-Generation Metropolis. In Robert A. Beauregard and Sophie Body-Gendrot (eds.), *The Urban Moment: Cosmopolitan Essays on the Late-20th-Century City*. London: Sage.

Maurrasse, David J. (2001). *Beyond the Campus: How Colleges and Universities Form Partnerships with Their Communities*. New York: Routledge.

McRoberts, Flynn, and Abdon Pallasch. (1999). "Neighbors Wary of New Arrivals." *Chicago Tribune* (December 28), pp. 1 and 12–13.

Mid-Continent Area Power Pool. (2001). *2001 Update to the 2000 Regional Plan: 2000 through 2009*. St. Paul, Minn.: Mid-Continent Area Power Pool.

Molotch, Harvey Luskin. (1972). *Managed Integration: Dilemmas of Doing Good in the City*. Berkeley: University of California Press.

Murray, Janet H. (1997). *Hamlet on the Holodeck: The Future of Narrative in Cyberspace*. New York: Free Press.

Myers, Dowell, and Alicia Kitsuse. (2000). "Constructing the Future in Planning." *Journal of Planning Education and Research*, 19, 221–231.

Myerson, George, and Yvonne Rydin. (1996). *The Language of Environment: A New Rhetoric*. London: UCL Press.

Newman, P., and J. Kenworthy. (1999). *Sustainability and Cities: Overcoming Automobile Dependence*. Washington, D.C.: Island Press.

Nieves, Evelyn. (2000). "Forget Washington. The Poor Cope Alone." *New York Times* (September 26), p. A1.

NSF Workshop on Urban Sustainability. (2000). *Toward a Comprehensive Geographic Perspective on Urban Sustainability*. New Brunswick, N.J.: Center for Urban Policy Research.

Nussbaum, Martha. (2001). *Upheavals of Thought: The Intelligence of Emotions*. New York: Cambridge University Press.

O'Connor, E. (1996). Telling Decisions: Organizational Decision-Making and Narrative Construction. In Z. Shapia (ed.), *Organizational Decision-Making*. New York: Cambridge University Press.

Pemberton, Gayle. (1992). Waiting for Godot on Jeffery Boulevard. In *The Hottest Water in Chicago: On Family, Race, Time, and American Culture* (pp. 176–193). Boston: Faber and Faber.

Peterson, Tarla Rae. (1997). *Sharing the Earth: The Rhetoric of Sustainable Development*. Columbia: University of South Carolina Press.

Petry, Ann. (1946). *The Street*. Boston: Houghton Mifflin.

Piore, Michael. (1995). *Beyond Individualism*. Cambridge, Mass.: Harvard University Press.

Plumwood, Val. (1998). Inequality, Ecojustice, and Ecological Rationality. In John S. Dryzek and David Schlosberg (eds.), *Debating the Earth: The Environmental Politics Reader* (pp. 559–583). New York: Oxford University Press.

———. (2002). *Environmental Culture: The Ecological Crisis of Reason*. London: Routledge.

Portelli, Alessandro. (1991). *The Death of Luigi Trastulli and Other Stories: Form and Meaning in Oral History*. Albany: State University of New York Press.

President's Council on Sustainable Development. (1996). *Sustainable America: A New Consensus for Prosperity, Opportunity, and a Healthy Environment*. Washington, D.C.: U.S. Government Printing Office.

Putnam, Robert D. (2000). *Bowling Alone: The Collapse and Revival of American Community*. New York: Simon & Schuster.

Rawls, John. (1971). *A Theory of Justice*. Cambridge, Mass.: Harvard University Press.

Reardon, Kenneth M. (2000). "Down on the River." *Planning*, 66, 20–23.

Reardon, Kenneth M., John Welsh, Brian Kreiswirth, and John Forester. (1993). "Participatory Action Research from the Inside: Community Development Practice in East St. Louis." *American Sociologist*, 24, 69–91.

Rein, Martin, and Donald Schon. (1977). Problem-Setting in Policy Research. In C. Weiss (ed.), *Using Social Research for Public Policy*. Lexington, Mass.: Lexington Books.

Revkin, Andrew C. (2001). "Global Warming Impasse is Broken." *New York Times* (November 11), p. A8.

Rintoul, Stuart. (1998). *The Wailing: A National Black Oral History*. Melbourne, Australia: Heinemann.

Robinson, Kim Stanley. (1990). *Pacific Edge*. New York: Tom Doherty Associates.

———. (1993, 1994, 1996). *Red Mars, Green Mars*, and *Blue Mars*. New York: Bantam Books.

Rotella, Carlo. (1998). *October Cities: The Redevelopment of Urban Literature*. Berkeley: University of California Press.

Roth, Henry. (Rpt. 1964, from 1934). *Call It Sleep*. New York: Avon.

Rusk, David. (1999). *Inside Game/Outside Game*. Washington, D.C.: Brookings Institution.

Ryden, Kent. (1993). *Mapping the Invisible Landscape*. Iowa City: University of Iowa Press.

Sachs, Wolfgang, Reinhard Loske, Manfred Linz, eds. (1998). *Greening the North: A Post-Industrial Blueprint for Ecology and Equity* (Timothy Nevill, trans.). London: Zed.

Sandercock, Leonie. (1998). *Towards Cosmopolis. Planning for Multicultural Cities*. Chichester, UK: Wiley.

———. (2000a). "Negotiating Fear and Desire: The Future of Planning in Multicultural Societies." Keynote paper, *Urban Futures Conference Proceedings. Urban Forum*, 11, 201–210.

———. (2000b). "Difference, Fear, and Habitus: A Reflection on Cities, Cultures, and Fear of Change." Keynote paper, Habitus Conference.

Schon, Donald A. (1984). *The Reflective Practitioner: How Professionals Think in Action*. New York: Basic Books.

Schudson, Michael. (1998). *The Good Citizen*. New York: Free Press.

Schwartz, Peter. (1996). *The Art of the Long View: Planning for the Future in an Uncertain World*. New York: Doubleday.

Scott, James C. (1998). *Seeing Like a State*. New Haven, Conn.: Yale University Press.

Sen, Amartya. (1999). *Development as Freedom*. New York: Knopf.

Sennett, Richard. (1970). *The Uses of Disorder*. New York: Knopf.

———. (1999). The Spaces of Democracy. In Robert A. Beauregard and Sophie Body-Gendrot (eds.), *The Urban Moment: Cosmopolitan Essays on the Late-20th Century City* (pp. 273–285). Thousand Oaks, Calif.: Sage.

Sharoff, Robert. (2000). "In Chicago, Two Bets on an Area's Revival." *New York Times* (April 2), p. 27.

Sloane, Paul E., Sr. (2001). "This Socialist Policy." *Louisville Courier-Journal* (August 13), p. A 6.

Smith, L. Tuhiwai. (1999). *Decolonizing Methodologies: Research and Indigenous Peoples*. London: Zed.

Soja, Edward. (1980). "The Socio-Spatial Dialectic." *Annals of the Association of American Geographers*, 70, 207–225.

———. (1989). *Postmodern Geographies: The Reassertion of Space in Critical Social Theory*. London: Verso.

———. (1996). *Thirdspace: Journeys to Los Angeles and Other Real-and-Imagined Places*. Malden, Mass.: Blackwell.

———. (2000). *Postmetropolis: Critical Studies of Cities and Regions*. Malden, Mass.: Blackwell.

Solnit, Rebecca. (2000). *Wanderlust: A History of Walking*. New York: Viking.

Stewart, Garrett. (1996). *Dear Reader: The Conscripted Audience in Nineteenth-Century British Fiction*. Baltimore, Md.: Johns Hopkins University Press.

Suarez, Ray. (1999). *The Old Neighborhood: What We Lost in the Great Suburban Migration, 1966–1999*. New York: Free Press.

Susskind, Lawrence, and Patrick Field. (1996). *Dealing with an Angry Public*. New York: Free Press.

Susskind, Lawrence, Sarah McKearnan, and Jennifer Thomas-Larmer, eds. (1999). *The Consensus Building Handbook*. Thousand Oaks, Calif.: Sage.

Taub, Richard P. (1988). *Community Capitalism: The South Shore Bank's Strategy for Neighborhood Revitalization*. Boston: Harvard Business School Press.

Taub, Richard P., D. Garth Taylor, and Jan D. Dunham. (1984). *Paths of Neighborhood Change: Race and Crime in Urban America*. Chicago: University of Chicago Press.

Taylor, Nigel. (1998). "Mistaken Interests and the Discourse Model of Planning." *Journal of the American Planning Association*, 64, 64–75.

Thomas, June Manning. (1994). "Planning History and the Black Urban Experience: Linkages and Contemporary Implications." *Journal of Planning Education and Research*, 14, 1–11.

Throgmorton, J. A. (1991). "The Rhetorics of Policy Analysis." *Policy Sciences*, 24, 153–179.

———. (1996). *Planning as Persuasive Storytelling: The Rhetorical Construction of Chicago's Electric Future*. Chicago: University of Chicago Press.

————. (2000). "On the Virtues of Skillful Meandering: Acting as a Skilled-Voice-in-the Flow of Persuasive Argumentation." *Journal of the American Planning Association, 66,* 367–379.

Todorov, Tzvetan. (1981). *Introduction to Poetics.* Minneapolis: University of Minnesota Press.

Torgerson, Douglas. (1999). *The Promise of Green Politics: Environmentalism and the Public Sphere.* Durham, N.C.: Duke University Press.

Urban Ecology. (1996). *Blueprint for a Sustainable Bay Area.* Oakland, Calif.: author.

Wackernagel, Mathis, and William Rees. (1996). *Our Ecological Footprint: Reducing Human Impact on the Earth.* Gabriola Island, British Columbia: New Society Publishers.

Walker, Caryn Faure. (2001). "Excavations: New Perspectives on the City and Cultural Space." *Public Art Journal,* 1.5 (April), 4–15.

Wasserman, Miriam. (2000). "Urban Sprawl." *Regional Review,* 10, 9–16.

Weir, Margaret. (1994). "Urban Poverty and Defensive Localism." *Dissent,* 41, 337–342.

White, Hayden. (1981). The Narrativization of Real Events. In W. J. T. Mitchell (ed.), *On Narrative.* Chicago: University of Chicago Press.

White, Richard. (1995). *The Organic Machine: The Remaking of the Columbia River.* New York: Hill & Wang.

Whyte, William H., Jr. (1958). "Urban Sprawl." *Fortune,* 57, 102–109ff.

Wideman, John Edgar. (1984). *Brothers and Keepers.* New York: Holt, Rinehart & Winston.

Wimsatt, W. K., Jr., and Monroe C. Beardsley. (1954). *The Verbal Icon: Studies in the Meaning of Poetry.* Lexington: University of Kentucky Press.

Winner, Langdon. (1986). *The Whale and the Reactor: A Search for Limits in an Age of High Technology.* Chicago: University of Chicago Press.

Wolch, Jennifer, Stephanie Pincetl, and Laura Pulido. (2002). Urban Nature and the Nature of Urbanism. In Michael. J. Dear (ed.), *From Chicago to L. A.: Making Sense of Urban Theory* (pp. 369–402). Thousand Oaks, Calif.: Sage.

World Commission on Environment and Development (1987). *Our Common Future.* Oxford: Oxford University Press.

Yanow, Dvora. (1995). "Built Space as Story: The Policy Stories That Buildings Tell." *Policy Studies Journal,* 23, 407–422.

Yardley, Jim. (2001). "New York's Sewage Was a Texas Town's Gold." *New York Times* (July 27), p. A 11.

Young, Iris Marion. (1990). *Justice and the Politics of Difference.* Princeton, N.J.: Princeton University Press.

————. (1996). Communication and the Other: Beyond Deliberative Democracy. In S. Benhabib (ed.), *Democracy and Difference* (pp. 120–135). Princeton, N.J.: Princeton University Press.

————. (2000). *Inclusion and Democracy.* Oxford: Oxford University Press.

Index